P9-CEM-008

Towing Icebergs, Falling Dominoes,

and Other Adventures in Applied Mathematics

ROBERT B. BANKS

Towing Icebergs, Falling Dominoes,

and Other Adventures in Applied Mathematics

Princeton University Press
Princeton, New Jersey

Published by Princeton University Press, 41 William Street,
Princeton, New Jersey 08540
In the United Kingdom: Princeton University Press,
Chichester, West Sussex

Library of Congress Cataloging-in-Publication Data

Banks, Robert.
Towing icebergs, falling dominoes, and other adventures in
applied mathematics / Robert B. Banks.
p. cm.
Includes bibliographical references and index.
ISBN 0-691-05948-9 (cl : alk. paper)
1. Mathematics—Popular works. I. Title.
QA93.B36 1998
510—dc21 98-4557

This book has been composed in Times Roman and Helvetica

Princeton University Press books are printed on
acid-free paper and meet the guidelines for permanence and
durability of the Committee on Production
Guidelines for Book Longevity on the
Council on Library Resources

http://pup.princeton.edu

Printed in the United States of America

10 9 8 7 6 5 4 3 2

To my mother,
Georgia Corley Banks,

and my brothers and sister,
Barney, Dick, and Joan

Contents

Preface

Mathematics is a field of knowledge that has many remarkable features. Perhaps the most remarkable feature of all is the one that enables humans to describe, measure, and evaluate the many aspects of endeavor that involve or affect human activity.

This book is a collection of topics that lend themselves to relatively simple mathematical analysis. The most important aspect of the book, however, is that all of these topics are concerned with phenomena, situations, events, and things that appear in all our lives. Some of these are *familiar* things—such as throwing baseballs, saving money, and jumping rope. Others are *imaginable* things—like icebergs being towed, federal debts being paid, and meteors crashing onto the earth.

These comments lead us to the following observation. It is indeed remarkable that the language of mathematics provides the mechanism we need to describe, accurately and concisely, the almost endless list of phenomena and things around us. How in the world would we be able to function without this beautiful language?

This is where mathematical models come in. In the book, simple models are developed to provide answers to questions. Some of these questions are almost trivial in nature; others are not. But all are questions that relate to human activity.

For example:

How far and how high do baseballs, golf balls and ski jumpers go and what are their velocities and flight times?

Commencing in the year 2000, how much money would the U.S. Congress need to budget annually in order to liquidate America's federal debt by, say, 2050?

The world's population is approaching six billion as we begin the new millennium. What will it be a hundred years from now?

"How Fast Can Runners Run?" is the title of the final chapter of the book. Well, it is predicted that by 2000, the record for the men's 100-meter dash will be 9.82 seconds and the record for the women's marathon will be 2 hours and 15 minutes.

The level of mathematics in this book ranges from elementary algebra and geometry to differential and integral calculus. Analytic geometry and spherical trigonometry are employed in the analysis of a number of the topics. Several chapters involve basic statistics and probability theory.

Where it is appropriate and feasible to do so, numerical examples are presented that involve various academic disciplines. Geography is featured in several examples and so are demography and economics. To a lesser extent, problems involving biology and physiology are examined.

I collected the topics presented in the book over quite a period of time. For about forty years, I was involved in teaching and research activities at various universities in America and abroad (England, Mexico, Thailand). As a professor of engineering, most of the courses I taught were in the areas of fluid mechanics and solid mechanics (statics, dynamics, mechanics of materials).

So this is a book about mathematics. It is certainly not a textbook, though it might be useful as a supplement to a text for high-school and university undergraduate students. Perhaps the book will be of most interest and usefulness to people who have already completed their formal educations. I believe there are many people in this category who really want to understand more about mathematics and the roles it plays in our high-technology world. I hope that study of the methods of solution of the problems—the easy ones as well as the difficult ones—will be helpful to these people.

And if, for these "continuing education students," time or other constraints do not make it possible always to understand every detail about the mathematical analysis of a particular problem, then at least the overall methodology of solution will be apparent and the main result available. For example:

For teams in the National Football League, the average period of oscillation between peak-to-peak seasonal win-loss performances is about 8.2 years and the average "turnaround time" is 2.1 years.

The shape of a deeply sagging flexible cable and the shape of the Gateway Arch in Saint Louis are the same. Both are *catenary* curves.

The shape of a child's jumping rope and the shape of a Darrieus wind turbine blade are identical. Both are called *troposkein* curves.

The velocity of the "wave crest" of a row of falling dominoes depends on the spacing of the dominoes. A typical velocity is 2.6 feet per second or about 1.77 miles per hour.

Finally, a word about the style I have used in writing the book. I decided to be a bit light-hearted in the analysis of some of the problems. So here and there, I have introduced small measures of humor and mirth. I hope that these efforts at levity are not too pathetic or corny. It just seems like a good idea not always to be entirely serious about everything.

Acknowledgments

I received a great deal of assistance from quite a few people during the preparation of this book. For their most appreciated reviews of various chapters of my manuscript, I would like to express my gratitude to the following persons: Marcel Berger, Harm de Blij, Gary Easton, Robert Heilbroner, Murray Klamkin, and James Murray.

In addition, several individuals provided me with all kinds of information related to the topics examined in this book. For this help, I want to thank Brad Andes, Mary Currie, Gilbert Felli, Sue Hansen, Sue Robertson, Gerard Saunier, and A. W. Vliegenthart.

Finally, I want to express my appreciation to my editor, Trevor Lipscombe, and his colleagues at Princeton University Press, for their great help in bringing order to the chaos of my manuscript.

In a special category of gratitude is the thanks I want to give to my wife, Gunta, for the immeasurable help she provided me throughout the long period I worked on this project. Without her wise advice and cheerful help, this literary iceberg could not have been towed anywhere.

Towing Icebergs,
Falling Dominoes,
and Other Adventures in
Applied Mathematics

1

Units and Dimensions and Mach Numbers

About twenty to twenty-five thousand years ago, an enormous meteor hit the earth in northern Arizona, approximately sixty kilometers southeast of the present-day city of Flagstaff. This meteor, composed mostly of iron, had a diameter of about 40 meters and a mass of around 263,000 metric tons. Its impact velocity was approximately 72,000 kilometers per hour or 20,000 meters per second. With this information, it is easy to determine that the kinetic energy of the meteor at the instant of collision was $e = (1/2)mU^2 = 5.26 \times 10^{16}$ joules. This is about 625 times more than the energy released by an ordinary atomic bomb.

This immense meteor struck the earth with such enormous force that it dug a crater 1,250 meters in diameter and 170 meters deep. More than 250 million metric tons of rock and dirt were displaced. The sound created by the impact must have been totally awesome.

As we shall see shortly, the velocity of sound in air is given by the equation $C = 20.07\sqrt{T}$, where C is the sonic velocity in meters per second and T is the absolute temperature of the air in degrees kelvin. For example, suppose that the air temperature is 68°F (fahrenheit) = 20°C (celsius) = 20°C + 273 = 293°K (kelvin); then the velocity of sound is $C = 20.07\sqrt{293} = 344$ m/s.

Had there been a city of Flagstaff when the meteor hit the earth, the people living there—60 kilometers away—would have heard the noise of the impact about 175 seconds after it occurred. Had there been a Los Angeles—260 kilometers to the west—the sound waves created by the collision would have reached there about 30 minutes later.

Over the years, scientists and engineers have devised several "numbers" that they use in mathematical analyses and computations involving the motion of objects moving through fluids such as water and air. By far the best known of these important numbers is the *Mach number*. It is highly likely, for example, that just about everyone has heard that the Concorde supersonic airliner, at cruising speed, has a Mach number of 2.0.

The Mach number, Ma, is defined as the velocity of an object moving through a fluid (e.g., water or air) divided by the velocity of sound in the same fluid. That is, $Ma = U/C$. In our meteor collision problem, $U = 20{,}000$ m/s and $C = 344$ m/s. Consequently, $Ma = 20{,}000/344 = 58$. This is a very large Mach number. The meteor was moving so fast just prior to impact that it created temperatures sufficiently high to ionize the air completely. This means that the molecules and atoms composing the air—mostly nitrogen and oxygen—were disintegrated into a gas called a "plasma." Ordinarily, even in high-speed aerodynamics, Mach numbers are much lower than the Mach number associated with the Arizona meteor. Typically, they are less than about 10. Never mind. For the moment, we simply want to present a definition of this quantity called the Mach number.

Units and Dimensions

In all fields of science and engineering, the subject of *units* and *dimensions* plays a very important role. In the physical and mathematical analyses of these fields, it is necessary to specify the fundamental dimensions of a measurement system and to define precisely the basic units to be used.

We should be careful to distinguish between the two quantities: units and dimensions. For example, *length* is a fundamental dimension; its units of measurement may be in feet, in miles, or in kilometers. *Time* is another fundamental dimension; its units may be expressed in seconds, in weeks, or in years; and so on.

For our purpose, there are five fundamental dimensions. These are the following:

mass, M, or force, F
length, L
time, T
electric current, A
temperature, θ

Sometimes it is preferable to use the dimension force, F, instead of mass, M. The two are easily interchanged because from Newton's equation, force = mass × acceleration, $F = M \times L/T^2$.

In our analysis, we consider the following systems of units:

the International System (SI) or metric system
the English or engineering system

For each of these, table 1.1 lists the proper units for the corresponding *fundamental dimensions*. For example, the SI or metric column indicates that the newton is the unit of force, the kilogram is the unit of mass, and the meter is the unit of length. Also, for each of the systems, the dimensions and units of several *derived quantities* are shown.

International System (SI) or Metric System

The metric system of units was originated in France following the French Revolution in the late eighteenth century. Being based on the units of meters, kilograms, and seconds, the metric system was referred to as the MKS system for many years. In 1960, it was replaced by what is called the International System (SI), which has been adopted by nearly all nations; eventually it will be used throughout the world.

Many scientists continue to use the centimeter-gram-second (CGS) system of units. This is the same as the SI system except

TABLE 1.1

Systems of units and corresponding dimensions

Quantities and dimensions	*SI or metric*	*English or engineering*
Fundamental		
Force F	Newton	Pound
Mass M	Kilogram	Slug
Length L	Meter	Foot
Time T	Second	Second
Current A	Ampere	Ampere
Temperature θ absolute	Kelvin, °K	Rankine, °R
Temperature θ relative	Celsius, °C	Fahrenheit, °F
Derived		
Velocity $U\ L/T$	Meters/second	Feet/second
Pressure $p\ F/L^2$	Newtons/meter^2	Pounds/foot^2
Density $\rho\ M/L^3$	Kilograms/meter^3	Slugs/foot^3
Energy $e\ FL$	Newton meters or joules	Foot pounds
Power $P\ FL/T$	Joules per second or watts	Foot/pounds/second

that the centimeter replaces the meter as the unit of length and the gram replaces the kilogram as the unit of mass.

A word about the temperature units indicated in table 1.1. In the SI system, absolute and relative temperatures are related by the equation °K = °C + 273.2. In addition, we have the relationship °C = (5/9)(°F − 32), where °F is degrees fahrenheit.

English or Engineering System

Most of the countries of the world have now adopted the SI system of units. Only in the United States, Great Britain, and some other English-speaking countries is the English/engineering system still being used. However, it is slowly being replaced by the much simpler and more logical SI system.

As table 1.1 indicates, the *pound* and the *slug* are the customary units for force (F) and mass (M). However, in Great Britain, the *poundal* is frequently taken as the unit of *force* (F). In this case, the unit of mass (M) is the pound.

In the English/engineering system, absolute and relative temperatures are related by the equation °R = °F + 459.7. In addition, we have °F = (9/5)°C + 32, where °C is degrees celsius.

Conversion of Units and Some Examples

A short list of numerical conversion factors is presented in table 1.2. Much longer lists are presented in many references. For example, a long table of conversion factors is given in Lide (1994).

TABLE 1.2

A short list of conversion factors between English or engineering and International System or metric

1 inch = 2.540 centimeters	1 pound = 0.4536 kilograms
1 foot = 30.48 centimeters	1 pound = 4.448 newtons
1 meter = 3.281 feet	1 slug = 32.2 pounds
1 mile = 5280 feet	1 kilogram = 2.205 pounds
1 mile = 1.609 kilometers	1 kilogram = 9.82 newtons
1 nautical mile = 6076.4 feet	1 calorie = 4.186 joules
1 nautical mile = 1852 meters	1 horsepower = 0.7457 kilowatts
°F = (9/5)°C + 32	°C = (5/9)(°F − 32)

PROBLEM 1. In the SI system of units, the acceleration due to gravity is $g = 9.82\ \text{m}/\text{s}^2$. What is its value in the English / engineering system?

$$g = 9.82\frac{\text{m}}{\text{s}^2} = 9.82\frac{(3.281\ \text{ft})}{\text{s}^2} = 32.2\frac{\text{ft}}{\text{s}^2}.$$

PROBLEM 2. In the SI system of units, the density of air is $\rho = 1.20\ \text{kg}/\text{m}^3$. What is its value in the English / engineering system?

$$\rho = 1.20\frac{\text{kg}}{\text{m}^3} = 1.20\frac{(2.205\ \text{lb})}{(3.281\ \text{ft})^3} = 0.0749\frac{\text{lb}}{\text{ft}^3},$$

$$\rho = 0.0749\frac{\left(\frac{1}{32.2}\ \text{slug}\right)}{\text{ft}^3} = 0.00233\frac{\text{slug}}{\text{ft}^3}.$$

PROBLEM 3. In the English / engineering system, the wind pressure on a tall building is $p = 45\ \text{lb}/\text{ft}^2$. What is its value in the SI system?

$$p = 45\frac{\text{lb}}{\text{ft}^2} = 45\frac{(4.448\ \text{newton})}{\left(\frac{1}{3.281}\ \text{meter}\right)^2},$$

$$p = 45(4.448)(3.281)^2\frac{\text{newton}}{\text{meter}^2},$$

$$p = 2{,}155\frac{\text{N}}{\text{m}^2} = 2{,}155\ \text{pascal}.$$

Prefixes for SI Units

These days we hear a lot about nanoseconds, megawatts, kilograms, and micrometers. We note that each of these SI units has a prefix. These prefixes give the precise size of the unit. A list of these prefixes, and their symbols and sizes, is given in table 1.3.

Dimensional Analysis

A topic closely related to the subject of units and dimensions, indeed one which is built entirely on the concept and theory of dimensions, is *dimensional analysis*. It is extremely important in

TABLE 1.3

Prefixes for SI units

Prefix	Symbol	Size: 10^k Value of k	Prefix	Symbol	Size: 10^k Value of k
deka	da	1	deci	d	−1
hecto	h	2	centi	c	−2
kilo	k	3	milli	m	−3
mega	M	6	micro	μ	−6
giga	G	9	nano	n	−9
tera	T	12	pico	p	−12
peta	P	15	femto	f	−15
exa	E	18	atto	a	−18

many areas of science and engineering, especially in the subjects of fluid mechanics and aerodynamics. We will not go much beyond a brief introduction to the topic. Numerous references are available: Barenblatt (1996), Ipsen (1960), and Langhaar (1951).

An Example: Flight of a Baseball and the Reynolds Number

To illustrate how dimensional analysis is used, we analyze a problem that is well known to nearly everyone: the flight of a baseball. In this case, a sphere of diameter D moves through a fluid (i.e., air) with velocity U. The fluid has density ρ_a and viscosity μ. The stitches and seam on the baseball create a rough surface that, like sandpaper, can be described by a certain roughness height ϵ.

The resistance force, F, that the fluid exerts on the sphere depends on a number of things. Mathematically, this dependence can be expressed in the following way:

$$F = f(D, U, \rho_a, \mu, \epsilon). \tag{1.1}$$

This relationship says that the resistance force F depends on—or, as a mathematician would say, is a function of—the diameter, D, and velocity, U, of the sphere, the density, ρ_a, and viscosity, μ, of the fluid through which the sphere is moving, and the roughness of the sphere, ϵ.

Altogether there are *six* variables in our problem; these are listed in equation (1.1). Collectively, these variables possess *three* of the fundamental dimensions: mass (M), length (L), and time (T). So the values of two important quantities in our dimensional analysis problem are $m = 6$ (number of physical variables) and $n = 3$ (number of fundamental dimensions).

The basic principle of dimensional analysis is contained in the following statement: Consider a system in which there are m independent dimensional variables that affect the system. Furthermore, there are n fundamental dimensions among these m quantities. Then it is possible to construct $(m - n)$ dimensionless parameters to relate these quantities functionally.

On this basis, in our problem, with $m = 6$ and $n = 3$, we can expect to construct $(m - n) = 6 - 3 = 3$ dimensionless parameters. Sure enough, if we were to go through the details of the entire dimensional analysis, we would obtain the following expression:

$$\frac{F}{\frac{1}{2}\rho_a A U^2} = f\left(\frac{\rho_a U D}{\mu}, \frac{\epsilon}{D}\right). \tag{1.2}$$

The quantity on the left-hand side of this equation expresses the resistance force; it is a dimensionless quantity. Likewise, the two quantities within the brackets on the right-hand side are also dimensionless quantities. Incidentally, when we say "dimensionless," we simply mean that the exponents of each of the fundamental dimensions in a particular parameter add up to zero. For practice, try checking the dimensions of the parameters of equation (1.2).

Equation (1.2) indicates that the term for the resistance force, $F(1/2)\rho_a AU^2$, is a function of the two quantities $\rho_a UD/\mu$ and

ϵ/D. We can rewrite this expression in the following way:

$$F = \frac{1}{2}\rho_a C_D A U^2, \tag{1.3}$$

in which $A = (\pi/4)D^2$ is the projected or shadow area of the sphere and C_D is the *drag coefficient*. It is clear that

$$C_D = f(Re, \epsilon/D), \tag{1.4}$$

where $Re = \rho_a UD/\mu$ is a quantity called the *Reynolds number*; the parameter ϵ/D is termed the relative roughness. In words, equation (1.4) says that the drag coefficient, C_D, depends on—or is a function of—the Reynolds number, *Re*, and the relative roughness, ϵ/D.

If the sphere is completely smooth—like a ping-pong ball, for example—then the roughness $\epsilon = 0$. In this case, the drag coefficient depends only on the Reynolds number. That is,

$$C_D = f(Re). \tag{1.5}$$

We note that the Reynolds number, $Re = \rho_a UD/\mu$, contains the viscosity, μ. Consequently, this important dimensionless number gives a measure of the importance of viscosity in a particular fluid flow phenomenon.

In later chapters, where we deal with baseballs, golf balls, and other objects moving through air, we shall take a close look at drag coefficients, Reynolds numbers, and the roughness caused by baseball seams and golf ball dimples. Why do we want to know about these things? Well, quite likely one of the main reasons is to be able to compute the trajectories—the flight paths—of baseballs and golf balls as they sail through the sky, in which case, as we shall see later on, it is absolutely imperative to have quantitative information about drag coefficients, lift coefficients, and the like.

However, our interest may go far beyond the task of simply calculating sporting ball flight paths. The same mathematics and physics are involved—though generally somewhat more complicated—if we want to compute the trajectories of projectiles, missiles, rockets, and yes, even ski jumpers.

Velocity of Sound in a Gas

When a sound wave passes through a gas—for example, air—the gas is slightly compressed momentarily by the wave. If we were to carry out a detailed analysis of this event, we would make the basic assumption that there is no gain or loss of heat into or out of the gas. In terms of thermodynamics, this says that the process is *adiabatic*. Utilizing this assumption and employing the so-called general gas law, we obtain the equation

$$C = \sqrt{\frac{\gamma R_*}{m} T}, \tag{1.6}$$

in which C is the velocity of sound in the gas, γ is the specific heat ratio of the gas, R_* is the universal gas constant, m is the molecular weight of the gas, and T is the absolute temperature. For air, $\gamma = 1.405$, $R_* = 8.314$ joules/°K mol, and $m = 29 \times 10^{-3}$ kg/mol. With these values, equation (1.6) becomes

$$C = 20.07\sqrt{T}, \tag{1.7}$$

which is the equation for the velocity of sound in air. It is interesting to note that the sonic velocity depends only on the temperature. For example, if $T = 20°\mathrm{C} = 293°\mathrm{K}$, then equation (1.7) gives $C = 344$ m/s, a result we obtained earlier in the chapter.

Velocity of Sound in a Liquid

Although it is usually assumed that liquids, including water, are incompressible, it turns out that they are, in fact, slightly compressible. If K is the so-called coefficient of compressibility of a liquid and ρ is its density, it can be shown that

$$C = \frac{1}{\sqrt{\rho K}}. \tag{1.8}$$

This is the equation for the velocity of sound in a liquid. For example, the value of K for sea water at 20°C is $K = 4.25 \times 10^{-10}$ m^2/newton and the density is $\rho = 1{,}025$ kg/m^3. If these numbers are substituted into equation (1.8), we obtain $C = 1{,}515$ m/s as the velocity of sound in sea water. At this same temperature, the velocity of sound in air is $C = 344$ m/s. Thus, the sonic velocity in the ocean is more than four times larger than it is in air. Most likely, whales and dolphins have known for quite a long time that vocal transmissions are much swifter *below* the surface.

External Forces in Fluid Flow Phenomena

We have seen that when the force of viscosity is the most important external force in a fluid flow, then dimensional analysis indicates that the Reynolds number, *Re*, is the important parameter involved in the problem. Likewise, if the force of compressibility is predominant, then a similar analysis predicts that the Mach number, *Ma*, is the crucial parameter of the phenomenon.

In the same way, if gravity is the major external force, then dimensional analysis would indicate that the dimensionless number called the *Froude number*, $Fr = U/\sqrt{gD}$, is the important parameter. Finally, if the major external force is due to surface tension, σ, then the *Weber number*, $We = \rho U^2 D/\sigma$ is the critical flow parameter. These are the important dimensionless numbers we mentioned at the beginning of the chapter.

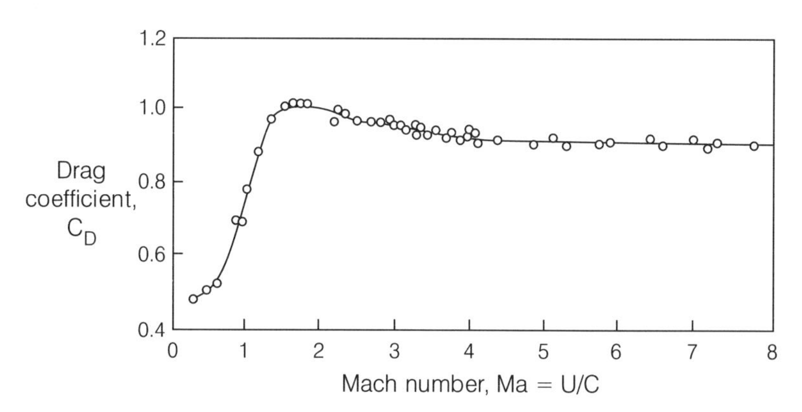

FIG. 1.1

Drag coefficient, C_D, versus Mach number, *Ma*, for smooth spheres. (From Barenblatt 1996.)

An Example: Flight of a Supersonic Sphere and the Mach Number

Suppose that a smooth sphere of diameter D moves at a very high velocity U through a compressible gas—for example, air. It is assumed that the effects of viscosity can be neglected.

In this case, the drag coefficient, C_D, depends only on the Mach number, $Ma = U/C$. This result, predicted by dimensional analysis, is confirmed by experimental results. A plot is presented in figure 1.1 of the drag coefficient versus the Mach number for the flow of air past a smooth sphere.

In the figure we note the following:

1. For values of Ma less than about 0.5 (i.e., in the subsonic region), the value of C_D has approximately the same value, $C_D = 0.48$, as in incompressible flow (e.g., the flight of a baseball or a golf ball).
2. For values of Ma from 0.5 to 1.5 (i.e., in the transonic region), there is a sharp increase in the value of the drag coefficient.
3. For a value of Ma equal to about 1.5 (i.e., in the supersonic region), the drag coefficient reaches a maximum value, $C_D = 1.02$.
4. For values of Ma greater than approximately 3.0 (i.e., in the supersonic-hypersonic region), the drag coefficient has a constant value of about 0.90.

The most important characteristic of supersonic flow—that is, a flow with Mach number larger than 1.0—is the appearance of shock waves. A reference that deals with this and numerous other topics of aerodynamics is Anderson (1991).

In conclusion, it should be mentioned that the Mach number is named after the noted Austrian physicist Ernst Mach (1838–1916). During the latter part of the nineteenth century, Mach carried out many studies on mechanics and aerodynamics, including extensive work involving wind tunnel experimentation. In addition to his numerous contributions in various areas of physical science, Mach is well known for his accomplishments in the fields of psychology and philosophy.

2

Alligator Eggs and the Federal Debt

About two hundred years ago, an English clergyman-economist named Thomas Malthus published a series of essays (1798, reprinted 1970) in which he contended that populations grow according to the law of geometric progression. That is, if a population of some locale has a certain magnitude at a particular moment, then that population will double itself at the end of a specified time period, and this periodic doubling of population will continue indefinitely. For example, if the population is $N = 1$ at time $t = 0$, then after each specified time period, t_2, the population will be 1, 2, 4, 8, 16, and so on.

This law of geometric progression or doubling is described by the equation

$$N = N_0 2^{t/t_2}, \tag{2.1}$$

in which N_0 is the value of N when time $t = 0$, and t_2 is what we call the "doubling time." This is essentially the mathematical expression formulated by Malthus.

We need not always be concerned with the doubling geometric progression of equation (2.1). We could just as well have a tripling, a quadrupling, or an "m-ling" geometric progression. In the general case, the equation of growth would be

$$N = N_0 m^{t/t_m}, \tag{2.2}$$

where t_m is the time period between successive generations. If $m = 4$, the growth sequence is 1, 4, 16, 64, 256, and so on. From equations (2.1) and (2.2) we obtain the relationship, utilizing natural logarithms $t_m = (\log_e m / \log_e 2)t_2$, where t_2 is the doubling time.

By way of example, we look into the potentially serious problem of alligator eggs in Florida. To simplify our analysis, we conveniently ignore virtually all concepts and principles of biology and ecology. We go directly to the heart of the problem.

Here it is: Suppose that somewhere in central Florida, at time $t = 0$, there is one alligator egg whose reproduction time is 30 days. Suppose also that after the 30 days this alligator egg turns into 16 alligator eggs. After another 30 days, each of the 16 becomes 16 more, so now we have 256 eggs. After another 30 days (now up to 90 days) we have a total of 4,096 alligator eggs.

Immediately we see that the equation describing the growth of the number of eggs is

$$N = (16)^{t/30}, \tag{2.3}$$

where, with reference to equation (2.2), $N_0 = 1$, $m = 16$, and $t_m = 30$. When $t = 120$ days we have $N = 65{,}536$ eggs and when $t = 180$ days there are 16,777,216 eggs, which is even more than the number of Floridians. At $t = 270$ days we have nearly 70 billion eggs—far more than enough to fill the enormous hangar at the Kennedy Space Center. Finally, when $t = 360$ days there are almost 300 trillion alligator eggs.

Now the area of the state of Florida is 58,664 square miles. If each egg occupies the space of a 2.5-inch cube, an easy computation shows that, after this period of almost one year, all Floridians will be up to their kneecaps in alligator eggs. Conclusion: geometric progressions lead to very large numbers in very short times.

For reasons of mathematical convenience, equation (2.1) can be written in the form

$$N = N_0 e^{at}, \tag{2.4}$$

which we call the exponential growth equation; the quantity a is termed the growth coefficient or interest rate. From equations (2.1) and (2.4) we establish that the growth coefficient (or interest rate) a and the doubling time t_2 are related by the expression

$$t_2 = \frac{\log_e 2}{a} = \frac{0.693}{a} \doteq \frac{70}{a(\%)}. \qquad (2.5)$$

For example, if the population growth rate of a nation is $a = 3.5\%$ per year, then the nation's population will double in about 20 years. If you save your money in a bank that compounds your interest earnings quite frequently (e.g., monthly, weekly, daily, instantaneously) at a rate $a = 7.0\%$ year, you will double your capital in around 10 years.

In his early essays, Malthus made very gloomy predictions about unbounded growth of populations, increasingly inadequate food supplies, and inevitable poverty and starvation. So-called Malthusian or exponential growth ignores all factors that provide restraints and limits to growth. These retarding factors, first proposed by Pierre Verhulst, a Belgian mathematician of the 1830s, led to the so-called logistic equation; we shall look at the logistic in the next chapter. In the meantime, all Floridians should rest easy about alligator eggs; there are, in fact, forces in all ecological and demographic settings that provide feedbacks to restrain or terminate growth.

At this point we back up briefly and write the following simple differential equation:

$$\frac{dN}{dt} = aN. \qquad (2.6)$$

This expression indicates that the rate at which a particular quantity grows, dN/dt, is directly proportional to the amount of the quantity, N, present at any moment. Essentially, this is a statement of Malthusian growth. This relationship explains why the number of alligator eggs increases so quickly to such a very large number: a kind of "rich-get-richer" scenario. We easily integrate equation (2.6) to obtain

$$N = N_0 e^{at},$$

which is equation (2.4). Also, from the above equations, the following relationships are established:

$$a = \frac{\log_e 2}{t_2} = \frac{\log_e m}{t_m}, \tag{2.7}$$

where, again, a is the growth coefficient, t_2 is the doubling time, and t_m is the reproduction time for an "m-ling" progression.

For the alligator eggs, with $m = 16$ and $t_m = 30$ days, we obtain $t_2 = 7.5$ days and $a = 0.09242$ per day. Substituting this value of a and $t = 360$ in equation (2.4) gives $N = 2.815 \times 10^{14}$, or about 300 trillion eggs, as we obtained before.

More amusement is provided by our next example: the growth of the federal debt of the United States. In 1945, at the conclusion of World War II, America's federal debt was just under $270 billion, and indeed it actually decreased to $250 billion in 1948. During the ensuing two decades the federal debt gradually increased; it had grown to around $373 billion by 1970. Subsequently, the debt began to show some signs of life; it rose to about $535 billion by 1975 and to $909 billion by 1980. And then, as would any self-respecting exponential function, the federal debt began to demonstrate real vitality; it started to exhibit the well-known alligator egg syndrome.

The amount of the federal debt is listed in table 2.1 at two-year intervals for the period 1970 to 1992; these amounts are displayed in graphical form in figure 2.1. Let us suppose that the data giving the growth of the federal debt, shown in figure 2.1 can be described by the exponential growth relationship of equation (2.4). Further, we want to use these data to determine the magnitudes of the initial value, N_0, and the growth coefficient, a, in equation (2.4). To accomplish this, we employ a statistical method called linear least-square analysis. A brief description of this method is given in the following section.

TABLE 2.1

Amounts of the federal debt of the United States, 1970 to 1992

Year	*t*	*N* *billion dollars*
1970	0	372.6
1972	2	427.8
1974	4	475.2
1976	6	621.6
1978	8	772.7
1980	10	908.7
1982	12	1,142.9
1984	14	1,573.0
1986	16	2,111.0
1988	18	2,586.9
1990	20	3,071.1
1992	22	4,077.5

Source: Data from U.S. Bureau of the Census (1994).

Least-Squares Analysis and Correlation Coefficients

In the analysis of experimental data, we frequently try to establish that the data are described by some kind of equation. In other words, we attempt to match or fit the data to a certain kind of mathematical curve. Many times we try to employ the simplest curve of all: the linear relationship of a straight line. In this instance, we employ the familiar equation $y = k_0 + k_1x$.

Recall that the intercept of this straight line with the y-axis is given by k_0 and the slope of the line is expressed by k_1. Since an infinite number of straight lines are available, we seek that line which provides the "best fit" to our data.

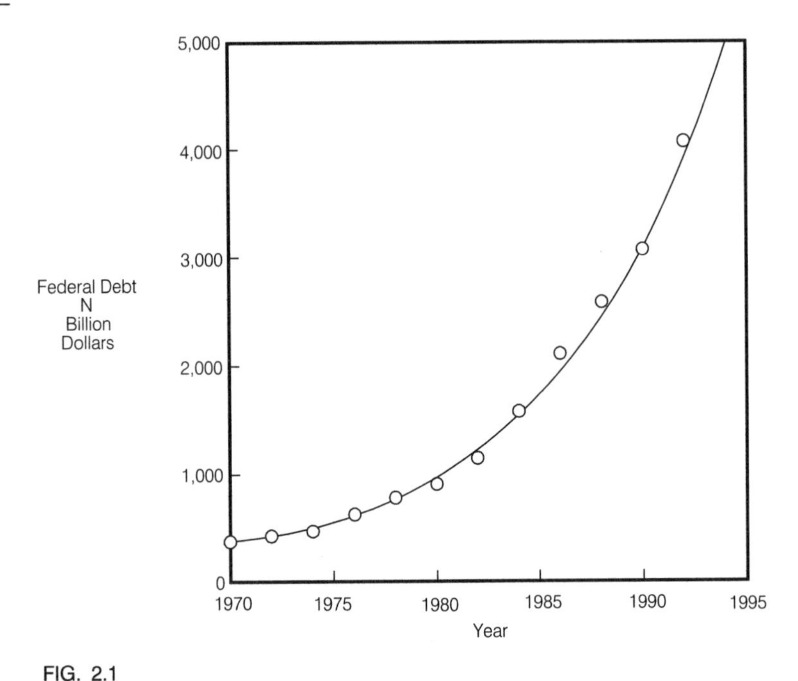

FIG. 2.1

Growth of the federal debt of the United States

Years ago, the famous German mathematician Carl Friedrich Gauss (1777–1855) formulated this problem and obtained its solution. Briefly, the following is the methodology he developed. It should be mentioned that the method is essentially the same for polynomials, for example, $y = k_0 + k_1x + k_2x^2$, and so on.

We let the quantity $\delta = y - (k_0 + k_1x)$ be the difference between the observed ordinate, y, and its computed value, $(k_0 + k_1x)$. Then, to avoid any confusion between positive and negative values, we square both sides and then obtain the sum of both sides. Letting S represent this sum, we have

$$S = \sum \delta^2 = \sum [y - (k_0 + k_1x)]^2.$$

We now say that the "best fit" of the data is obtained when the magnitude of S is a minimum, that is, the "least-squares" value. Accordingly, to obtain this minimum value, we determine the derivatives of S with respect to k_0 and k_1, and set these equal to zero. This provides two equations to compute the "best" values of these two constants. The entire procedure is called the *least-squares analysis*.

In carrying out this methodology, our analysis generates a statistical index called the *correlation coefficient*, r. This quantity simply measures the "strength" of the linear relationship between x and y. The closer the value of r is to unity, the stronger is the linear relationship. Values of r close to zero indicate a weak linear relationship between x and y.

Virtually all scientific calculators are programmed to carry out the least-squares analysis and correlation coefficient calculations. Most books dealing with elementary statistics discuss these topics. Two recommended references are Mosteller et al. (1983) and Sellers et al. (1992).

Back to the Federal Debt and Alligator Eggs

We use equation (2.4) as the mathematical framework for our federal debt problem. Taking the logarithms of both sides of that equation gives the expression

$$\log_e N = \log_e N_0 + at, \tag{2.8}$$

which has the simple mathematical form $y = k_0 + k_1 x$. A least-squares analysis of the data of table 2.1 yields $N_0 = \$326$ billion and $a = 0.113$ per year. The correlation coefficient is $r = 0.9964$. We note that there is a very high degree of correlation between the variables.

If it is assumed that the federal debt continues to increase according to the trend described by equation (2.4), the amount of the debt will be over \$9 trillion by the end of the century. This is serious money and many economists say it reflects a very serious situation. We escape to less troublesome matters with a return to our alligator problem.

We left off with $N = 281.5$ trillion eggs at $t = 360$ days. They cover the entire state of Florida to a depth of 18 inches. Good news! It is decided to launch immediately a massive program to dump all these eggs into the Atlantic Ocean or the Gulf of Mexico. We assume the eggs simply sink and do not turn into alligators. However, throughout the clean-up program the eggs continue to increase at the same rate as before. Question: How many eggs do we have to dump each day to get rid of all of them in a specified period of time?

To answer this question, we need the following equation:

$$\frac{dN}{dt} = aN - h, \tag{2.9}$$

which is the same as equation (2.6) except that we have subtracted the quantity h. This represents the number of alligator eggs dumped into the ocean every day. A demographer would call this quantity the emigration rate, a bioeconomist the harvesting rate.

First we rewrite equation (2.9) in the following integral form:

$$\int_{N_*}^{0} \frac{dN}{aN - h} = \int_{0}^{t_e} dt. \tag{2.10}$$

The quantities appearing in the lower limits of the integrals indicate that when $t = 0$ (the day the cleanup program starts), the number of alligator eggs is $N_* = 281.5 \times 10^{12}$. The upper limits of the integrals stipulate that $N = 0$ (our goal) when $t = t_e$, the "extinction" time. Next, we integrate equation (2.10) and with a little algebra obtain the equation

$$t_e = \frac{1}{a} \log_e\left(\frac{h}{h - aN_*}\right). \tag{2.11}$$

All we need do now is substitute the numerical value of the removal rate, h, into this equation and compute the extinction time, t_e. We note, however, that h must be at least as large as the product $AN_* = (0.09242)(281.5 \times 10^{12}) = 26.02 \times 10^{12}$ per day. If not, the number of eggs wlll increase faster than we can dump them.

We try $h = 30$ trillion eggs per day. The answer provided by equation (2.11) is $t_e = 21.8$ days. If $h = 50$ trillion per day, then

$t_e = 8.0$ days and if $h = 100$ trillion per day, we get $t_e = 3.3$ days. A formidable task, whatever the schedule.

PROBLEM. Back to the federal debt. Please examine the following: If the federal debt of the United States is halted at a magnitude of, say, \$6 trillion in 2000 and a debt retirement program is commenced immediately, how much will Congress need to appropriate each year, with an interest rate $a = 0.06$, to liquidate the debt in (a) 25 years and (b) 50 years?

Answers. (a) \$463 billion and (b) \$379 billion.

3

Controlling Growth and Perceiving Spread

We saw in the previous chapter that things which multiply according to geometric progression, or equivalently, exponential growth, can increase to very large amounts in relatively short times. Good examples are the number of alligator eggs, the amount of the federal debt, and your savings account earning compound interest.

Something always spoils the fun and here it is: Things cannot grow exponentially forever. A mathematically inclined proud father keeps a record of the height of his son during the period the son is from one to five years old. He then comes up with a great formula: $H = 28.2e^{0.09t}$, where H is the height of his son in inches and t is his age in years. From this, the father calculates that his son will be slightly over nine feet tall by the time he is fifteen. Great basketball player.

The same thing with the federal debt; at least, we hope so. We developed the equation $N = 0.326e^{0.113t}$, in which N is the amount of the federal debt in trillions of dollars and t is time in years measured from 1970. On this basis, the federal debt will be over 9 trillion dollars in the year 2000 and approaching 163 trillion by the year 2025.

Well, a few years ago it became obvious that this ridiculous situation could not continue. The United States Congress was fearful that the nation's entire income would have to be used to pay the interest on the public debt. So it legislated the following:

> Be it the demand of Congress that commencing with fiscal year 1993, authorized annual budgets must assure that the federal debt will never, absolutely never, exceed $6,000 billion.
>
> Be it the further demand of Congress that you and I, as expert consultants, work out the necessary details for programming this debt reduction scheme.

In 1838 a very clever Belgian mathematician named Pierre Francois Verhulst published the results of studies he had carried out involving population growths of Belgium, France, England, and Russia. In essence, Verhulst modified the Malthus equation by simply attaching the bracketed term in the following equation:

$$\frac{dN}{dt} = aN\left(1 - \frac{N}{N_*}\right), \tag{3.1}$$

where N_* is the so-called equilibrium value or carrying capacity. This expression says that if N is small in comparison with N_* then the growth rate, dN/dt, is approximately the same as that of exponential growth. On the other hand, if N is large or indeed equal to N_* then the growth rate becomes zero and so N can increase no further.

What an easy and logical way to get around the foolishness of Malthusian exponential growth. The solution to equation (3.1) is

$$N = \frac{N_*}{1 + (N_*/N_0 - 1)e^{-at}}. \tag{3.2}$$

This equation is termed the Verhulst equation or, more frequently, the logistic equation.

We decide to use equation (3.2) as the mathematical mechanism to bring the federal debt under control. We start with the following information: as demanded by Congress, $N_* = 6{,}000$ billion and $t = 0$ in 1993, and from chapter 2 $a = 0.113$ per year and $N_0 = 4{,}385$ billion. Substituting these numbers into equation (3.2) and computing N for various values of time t, we obtain the

TABLE 3.1

Amounts of the federal debt of the United States

Year	t	*N: debt controlled billion dollars*	*N: debt uncontrolled billion dollars*
1993	0	4,385	4,385
1994	1	4,515	4,909
1995	2	4,638	5,497
1996	3	4,753	6,154
1997	4	4,861	6,890
1998	5	4,961	7,715
1999	6	5,055	8,638
2000	7	5,144	9,671
2010	17	5,693	29,938
2025	32	5,941	163,063

amounts shown as "N: debt controlled" in table 3.1. The column with the heading "N: debt uncontrolled" lists the amounts computed from the exponential growth equation.

It is clear that our efforts to reduce the federal debt paid off. As seen in the table, by the year 2000 the "controlled" federal debt is calculated to be $5,144 billion instead of the "uncontrolled" $9,671 billion it would otherwise have been.

Since our task for the U.S. Congress has been completed, let us journey to the serene world of botany and look at sunflowers. Quite a few years ago, two American botanists, Reed and Holland, carried out experiments involving the growth of sunflowers. Specifically, they measured the average heights of sunflower plants each week over a period of many weeks. Their experiments have become something of a classic; their data have been analyzed by Feller (1940), Lotka (1956), and numerous other investigators.

These sunflower growth data are shown in figure 3.1. For the present, we ignore the solid curve in the figure. The assumption is made that the Verhulst or logistic equation provides a suitable mathematical framework for the description of sunflower growth.

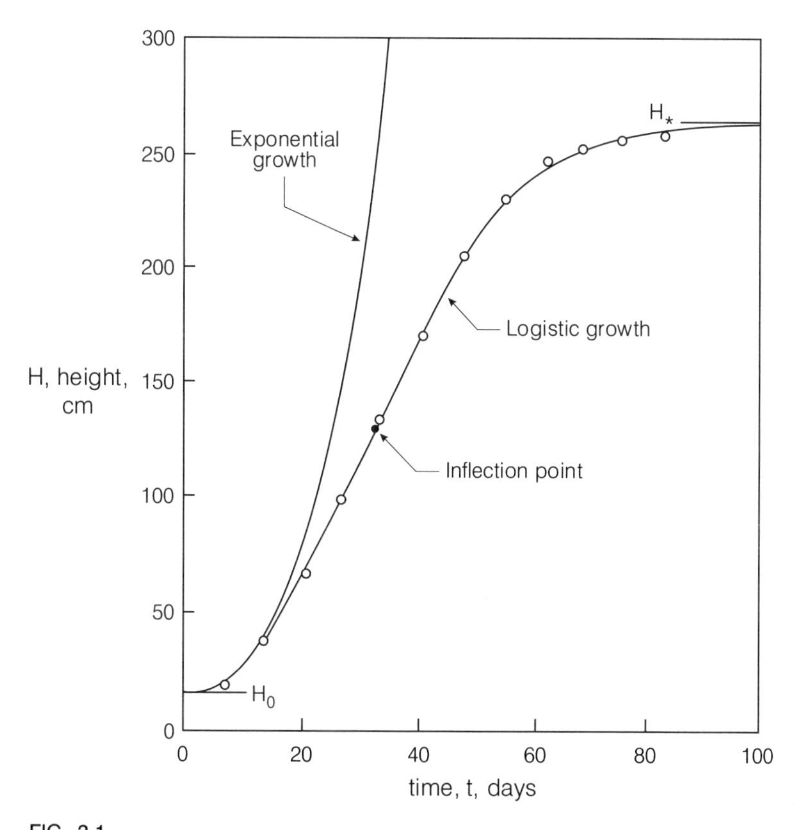

FIG. 3.1

Growth of sunflower plants. Data of Reed and Holland. (From Lotka 1956.)

That is,

$$H = \frac{H_*}{1 + (H_*/H_0 - 1)e^{-at}}, \tag{3.3}$$

where H is height in centimeters and t is time in days. Initial height and equilibrium height are H_0 and H_*, respectively. From the observed data, plotted in the figure, we determine the following: $a = 0.0876$ per day, $H_0 = 12.4$ cm, and $H_* = 261.1$ cm. Substitution of these numbers into equation (3.3) yields the solid curve in figure 3.1. It is reasonable to conclude that sunflower plant height is a good example of logistic growth.

We notice in figure 3.1 that initially the rate of growth, that is, the slope of the curve, is relatively flat. However, with increasing time the curve becomes steeper and steeper until it reaches the so-called inflection point, where the rate of growth has its maximum value. Then, as more time passes, the slope becomes flatter and flatter until eventually it becomes zero.

PROBLEM. Show that the inflection point of the logistic equation has the following coordinates and slope:

$$t_i = \frac{1}{a} \log_e \left(\frac{H_*}{H_0} - 1 \right); \qquad H_i = \frac{1}{2} H_*; \qquad \left(\frac{dH}{dt} \right)_i = \frac{1}{4} a H_* . \tag{3.4}$$

For our growing sunflowers, the inflection point occurs at $t_i = 34.2$ days. At that time, the height will be $H_i = 130.6$ cm, which is exactly one-half the ultimate height, and the flowers will be growing at the rate $(dH/dt)_i = 5.72$ cm/day.

Shown also in figure 3.1 is the curve corresponding to exponential growth. In this case we have, of course,

$$H = H_0 e^{at}, \tag{3.5}$$

where $a = 0.0876$ per day and $H_0 = 12.4$ cm. Instead of leveling off at $H_* = 261.1$ cm, our exponentially growing sunflower continues upward at a very alarming rate. For example, when $t = 93.4$ days, the height of the plant is $H = 44{,}335$ cm $= 443.35$ m $=$ 1,454 ft, which is the height of the Sears Tower in Chicago. And it continues to grow!

We have had a glimpse of the subject of *controlling growth*. Now we take a look at another topic we shall call *perceiving spread*.

This subject has to do with the way things are diffused or dispersed or disseminated or spread. Some examples of "things" are news, rumors, jokes, fads, ideas, innovations, and technologies. More "things": diseases, heat, pollution, traffic, insects, and people. Broadly speaking we could put these things into two categories: spreading of information and spreading of mass or energy. More importantly, in general, spreading takes place in both time and space.

Most problems involving dispersion or spreading are quite complicated: atmospheric pollution, expanding cities, spreading of HIV-AIDS, killer bee invasions, forest blights, and so on. We shall begin and end with a simple example.

We have a community, say a town or a university campus, with maybe 10,000 people. At time zero in our example, one person has a joke or a nice item suitable for gossip. That person tells the joke or gossip item to another person, so now two people know. These two each tell two more people and hence now there are four knowledgeable people.

You can see it coming: our geometric progression. At any given moment, N people have heard the joke or rumor and are spreading it; these are the "infectives." At that same moment, $N_* - N$ people have not heard the joke or rumor; these are the "susceptibles." The quantity N_* is the total number of people in the community; in our case, $N_* = 10{,}000$.

After a certain period of time, infectives start telling the joke to those who are also infectives; they have already heard the joke and, of course, are also spreading it around. After an additional period of time, susceptibles are increasingly hard to find; virtually everyone has heard the joke.

If we were to carry out a so-called stochastic analysis of this problem we would obtain a "difference equation" that would tell us how many infectives there are at any time t. However, if the assumption is made that the total population, N_*, is very large we can replace the stochastic analysis with a deterministic analysis and acquire a "differential equation." In this case we obtain the relationship

$$\frac{N}{dt} = aN - bN^2, \tag{3.6}$$

in which N is the number of infectives (those who have already heard the joke), a is the growth or spreading coefficient, and b is what we call a crowding coefficient. The number of susceptibles (those who have not yet heard) is, of course, $N_* - N$. Defining the crowding coefficient as $b = a/N_*$, equation (3.6) becomes

$$\frac{dN}{dt} = aN\left(1 - \frac{N}{N_*}\right), \tag{3.7}$$

which is the differential equation for logistic growth. Letting N_0 be the value of N when $t = 0$, the solution to equation (3.7) is given by equation (3.2).

An amazing result: The equation that describes the spreading of a joke, a rumor, or a simple epidemic also describes the growth of sunflower plants and controlled federal debts. A number of topics dealing with growing and spreading are presented in interesting books by Bartholomew (1981), Lighthill (1978), and Thompson (1963).

4

Little Things Falling from the Sky

There are always things falling from up there somewhere: smoke particles, dust, and volcanic ash; rain, snow, sleet, hail, and other objects that do not deter the postal service; leaves, bombs, Wall Street bankers, and skydivers. Some of us have been targets of seagulls as we walk along the beach. Others have caught baseballs sailing out of Wrigley Field as the Chicago Cubs hit yet another home run. Still others, enjoying coffee and newspaper in the kitchen, are distracted when a small meteor crashes through the ceiling. Isaac Newton, impacted by apples, devised the universal law of gravitation. Even the scriptures (Exodus 16: 14–36) describe manna from heaven—bread descending from the sky.

With that as a prologue, we begin our study of falling objects with a journey to Pisa, in Italy, about four hundred years ago.

Galileo and Newton and Falling Objects

Around the end of the sixteenth century, Galileo Galilei (1564–1642), an Italian astronomer and mathematician, carried out lengthy studies of the motion of projectiles. As part of those studies, according to the popular story, he dropped objects of various weights and sizes from the top of the Leaning Tower of Pisa and observed their times of fall.

Galileo concluded that the velocity of a falling object depends only on the distance of fall, not on its size or weight. He put to rest, or thought he did, the then-held view that the larger and heavier an object is, the faster it falls. He was partly right, partly wrong.

The same year that Galileo died, Isaac Newton (1643–1727) was born. The famous laws of motion of this English genius cleared up most of the confusion concerning Galileo's results. His second law was especially helpful. This law states that the summation of all the forces acting on an object is equal to the mass of the object times its acceleration.

For an object falling through air, this relationship gives

$$\rho_s g V - \rho_a g V - F_D = \rho_s V a. \tag{4.1}$$

The first term on the left expresses the weight of the object, the second term the buoyant force, and the third term the drag force. In this equation, ρ_s is the density of the material composing the falling object, ρ_a is the density of air, g is the gravitational force per unit mass, and V is the volume. The acceleration of the object is a. We assume that ρ_a is very small compared to ρ_s. In this case, the second term of equation (4.1) can be dropped.

By definition, the acceleration, $a = dU/dt$ and $U = dy/dt$, where y is the distance of fall and U is the velocity. So the acceleration can be written as $a = U(dU/dy)$, and equation (4.1) becomes

$$\rho_s g V - F_D = \rho_s V U \frac{dU}{dy}. \tag{4.2}$$

Case 1. Object Falling in a Vacuum

Suppose there is no air. In this case, there cannot be any drag force and so $F_D = 0$. Accordingly, equation (4.2) simplifies to

$$g = U \frac{dU}{dy}, \tag{4.3}$$

which can be written in the form

$$\int_0^U U\,dU = g\int_0^y dy. \tag{4.4}$$

The lower limits of the integrals indicate that $U = 0$ when $y = 0$; that is, we simply drop, not throw, the object. Integrating equation (4.4) and solving for U gives

$$U = \sqrt{2gy}\,. \tag{4.5}$$

From this equation and the relationships, $U = dy/dt$ and $a = dU/dt$, we also obtain

$$y = \tfrac{1}{2}gt^2; \qquad U = gt; \qquad U = \sqrt{2gy}\,; \qquad a = g. \tag{4.6}$$

Now this is where Galileo was right. In the absence of air, that is, in a perfect vacuum, the velocity depends only on the distance of fall.

Case 2. Object Falling in Air

Fortunately or unfortunately, air exerts a force on a moving object. It allows airplanes to fly, hailstones to slow down, and skydivers to survive. It also increases fuel consumption of cars and trains, knocks down bridges and buildings, and creates destructive waves on oceans and shores.

In any event, and for better or worse, air resistance cannot be ignored and so the drag force, F_D, is not zero. A so-called dimensional analysis shows that the drag force exerted on an object is

$$F_D = \tfrac{1}{2}\rho_a C_D A U^2, \tag{4.7}$$

in which C_D is a dimensionless drag coefficient, A is the shadow or projected area of the falling object, and U is the velocity.

Spheres Falling through Air

For simplicity, suppose that the falling object is a solid sphere of diameter D. Then the projected area is $A = \pi D^2/4$ and the volume is $V = \pi D^3/6$. Consequently, equation (4.2) becomes

$$\rho_s g \frac{\pi}{6} D^3 - \frac{1}{2} \rho_a C_D \frac{\pi}{4} D^2 U^2 = \rho_s \frac{\pi}{6} D^3 U \frac{dU}{dy}, \tag{4.8}$$

which simplifies to

$$U \frac{dU}{dy} = g - \frac{3\rho_a C_D}{4\rho_s D} U^2. \tag{4.9}$$

Because of the drag force, after a certain distance of fall the sphere attains a constant velocity, the so-called terminal velocity. Accordingly the acceleration is zero and so $dU/dy = 0$ in equation (4.9). Solving for U_* we get

$$U_* = \sqrt{\frac{4\rho_s g D}{3\rho_a C_D}}, \tag{4.10}$$

where U_* is the terminal velocity of the sphere.

It turns out that the drag coefficient, C_D, depends on the Reynolds number, $Re = \rho_a UD/\mu_a$, where μ_a is the viscosity of air. A plot of the C_D versus Re relationship for a smooth sphere is shown in Figure 4.1 When the Reynolds number is small ($Re < 1$), the relationship is $C_D = 24/Re$. Consequently, equation (4.10) reduces to

$$U_* = \frac{\rho_s g D^2}{18\mu_a}. \tag{4.11}$$

This expression, called Stokes' law, gives the terminal velocity, U_*, for very small spheres. For example, for water droplets falling in air, the limiting value, $Re = 1$, corresponds to $U_* = 19$ cm/s and D = 0.008 cm. We would call this mist or drizzle.

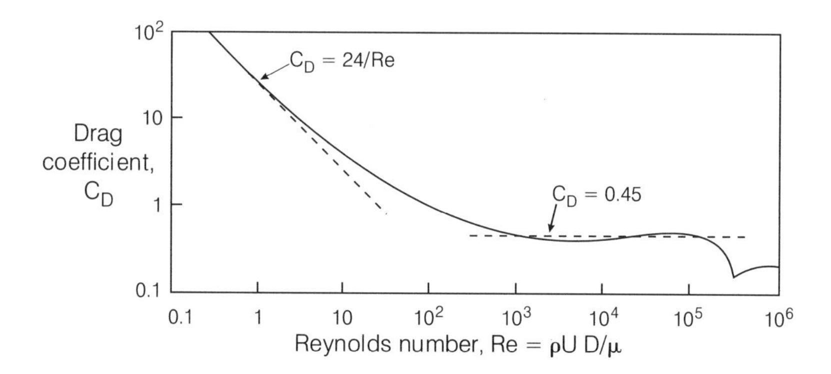

FIG. 4.1

Drag coefficient curve for a smooth sphere. (From Streeter and Wylie 1985.)

When the Reynolds number is large (Re is in the range from 1,000 to 100,000), $C_D \doteq$ constant $\doteq 0.45$. In this case, equation (4.10) describes the terminal velocity, U_*.

Here is a good question. If an object, for example, a sphere, is dropped from a certain height, through what distance must it fall to attain a terminal velocity? To keep the problem simple, it is necessary to assume that C_D is constant. With this fairly reasonable assumption, integration of equation (4.9) and use of equation (4.10) gives the expression

$$U = U_* \sqrt{1 - e^{-y/y_0}}, \tag{4.12}$$

where $y_0 = V_*^2/2g$. This quantity, y_0, might be called a "distance constant." This equation says that $U = 0$ when $y = 0$, as required. It also says that when $y = y_0$, the velocity U is 79.5% of the terminal velocity U_*; when $y = 2y_0$, U is 93.0% of U_*; and when $y = 3y_0$, U is 97.5% of U_*. So we can say that the terminal velocity is attained when $y = (3 \text{ or } 4)y_0$.

Using equation (4.12) and the relationship $U = dy/dt$, we also obtain

$$y = y_0 \log_e \cosh^2\left(\frac{gt}{U_*}\right), \tag{4.13}$$

and

$$U = U_* \tanh\left(\frac{gt}{U_*}\right); \qquad a = g\,\mathrm{sech}^2\left(\frac{gt}{U_*}\right). \tag{4.14}$$

These expressions contain several of the so-called hyperbolic functions (e.g., hyperbolic cosine, hyperbolic tangent). Though less familiar than the well-known circular functions (e.g., sine, cosine, tangent), they are every bit as useful.

Corresponding to equations (4.12) to (4.14) are equations (4.5) and (4.6) for an object falling in a vacuum. In equation (4.12), if the fall distance is "small," we can use the series approximation

$$e^{-y/y_0} \doteq 1 - y/y_0 \tag{4.15}$$

to obtain $U = \sqrt{2gy}$. For small values of time, equations (4.13) and (4.14) revert to the corresponding expressions of equation (4.6). In other words, for small distances and times, the sphere behaves as though it were falling in a vacuum.

Finally, for use in computations later on, it is convenient to use equations (4.12) and (4.14) to obtain the following expression for the impact time:

$$t = \frac{2y_0}{U_*} \operatorname{arctanh}\sqrt{1 - e^{-y/y_0}}. \tag{4.16}$$

Dropping Spheres from the Top of Tall Structures

The Leaning Tower of Pisa was constructed during the period from 1174 to 1271. It is a circular eight-story structure with a total height of fifty-five meters. The base has a diameter of sixteen meters with walls over four meters thick at the ground level and about half that amount near the top. In 1829 the structure was already over four meters off the vertical and by 1910 it was off by about five meters. The tilt angle at present is 5° 15′; the foundation has been strengthened to prevent further leaning.

Allegedly, it was from the top of this tower, in the year 1590 or so, that Galileo conducted his famous experiments on dropping objects. In their interesting book *Engineering in History*, Kirby et al. (1990) indicate that these experiments probably never happened. Maybe not, but it makes a good story.

In any event, we assemble our research group on the seventh level of the tower ($y = 45$ m), about four hundred years later, to carry out some tests. We have four spheres to drop: two are made of iron and two of rock. Two of the spheres are the size of baseballs and two the size of golf balls. The various constants, parameters, and (computed) results are shown in table 4.1.

As indicated in the bottom row of the table, the impact velocity of a sphere dropped a distance $y = 45$ m, in a vacuum is

TABLE 4.1

Summary of results of sphere-dropping experiments from the Leaning Tower of Pisa and the World Trade Center

Sphere number	*Density* ρ_s, *g / cm*^{3}	*Diameter D, cm*	*Mass m, g*	*Distance constant* y_0, *m*	*Terminal velocity* U_*, *m / s*
1	7.5 (iron)	7.5	1657	695	116.78
2	7.5 (iron)	4.2	291	389	87.39
3	2.5 (rock)	7.5	552	232	67.42
4	2.5 (rock)	4.2	97	130	50.45

Sphere number	*Tower of Pisa Impact velocity* U_i, *m / s*	$y = 45$ *m Impact time* t_i, *s*	*World Trade Center Impact velocity* U_i, *m / s*	$y = 416$ *m Impact time* t_i, *s*
1	29.25	3.046	78.37	9.674
2	28.88	3.057	70.82	10.044
3	28.33	3.083	61.55	10.634
4	27.32	3.125	49.41	11.766
Vacuum	29.73	3.027	90.39	9.205

Note: $\rho_a = 1.2 \times 10^{-3}$ g/cm^3; $g = 982$ cm/s^2; $C_D = 0.45$

$U_i = 29.73$ m/s and the impact time is $t_i = 3.027$ s. The impact velocities of the four spheres, computed from equation (4.12), and the impact times, calculated from equation (4.16), are shown in the lower left columns of the table.

We note that the impact times of the four spheres dropped from the tower are very nearly the same and indeed are only slightly larger than the impact time in a vacuum. On the basis of these playlike results, perhaps we can understand why Galileo reached the conclusion he did. Even in our experiments, with a fairly wide range of sizes and weights, impact velocities and impact times are almost constant. Our main conclusion is that the 45-meter height of the tower of Pisa was simply insufficient for Galileo to observe differences in velocities and times.

So our research group moves on to New York to repeat the tests from the top of the World Trade Center, height $y = 416$ m. Our (calculated) measurements are shown in the lower right columns of table 4.1. In this series of experiments we do have measurable differences of impact velocities and times. So now we can conclude that because of the drag force of air, fall velocities of objects do indeed depend on size and weight.

How to Design a Parachute

Now for an exciting project. What we want to do is determine how big a parachute must be to bring a person or a cargo safely from an airplane to the ground. We start with the relationship

$$W - F_D = ma, \tag{4.17}$$

where W is the weight of the person or cargo, including the chute, F_D is the drag force, $m = W/g$ is the mass of the person or cargo, and a is the acceleration. As before, we have

$$W - \frac{1}{2}\rho_a C_D A U^2 = \frac{W}{g} U \frac{dU}{dy}. \tag{4.18}$$

For our parachute design we are interested only in the terminal velocity, U_*. Setting $dU/dy = 0$ in equation (4.18) and solving for U_* gives

$$U_* = \sqrt{\frac{2W}{\rho_a C_D A}}. \tag{4.19}$$

What we want to determine, of course, is the size of the parachute, that is, the projected area, A. So we need to specify W, ρ_a, C_D, and U_*. For a change, let us use the English system of units instead of the International System (SI).

a. *Weight*, W. We select a design weight $W = 200$ lb.

b. *Air density*, ρ_a. In the English (or engineering) system, $\rho_a = 0.0024$ slugs/ft^3 at pressure and temperature of $p_0 = 14.7$ lb/in^2 and $T = 60°$ F.

c. *Drag coefficient*, C_D. Extensive information concerning aerodynamic drag coefficients is given by Hoerner (1965). An open parachute has the shape of a hollow hemisphere. For this shape, we have the following information: $C_D = 1.40$ when the hollow open side faces the air flow, and $C_D = 0.42$ when the closed solid side faces the air flow.

d. *Impact velocity*, U_*. With what velocity can a person safely land on the ground? Like jumping off a chair? Too slow. Like jumping from a second-story window? Too fast. How about off a ladder six feet high? Seems about right. So, $U_* = \sqrt{2gh} = \sqrt{2(32.2)(6.0)} =$ 19.66 ft/s.

Substituting these numbers into equation (4.19) gives $A = 308$ ft^2, and so the diameter of the parachute is $D = 19.81$ ft. We shall use $D = 20$ ft. With this size of parachute the terminal velocity is $U_* = 19.5$ ft/s or about 13.3 mi/hr.

Wherein We Go Skydiving

We conclude this chapter about falling objects with the following assignment. It is entirely optional.

With the parachute you have now designed and fabricated, properly packed and attached to your body, please step out of the airplane door. Do not open your parachute yet. Please assume the basic free-fall position: face to earth, horizontal, knees slightly bent, arms angled at about 45° in the horizontal plane. In this position, according to data obtained in a wind tunnel and published by Hoerner (1965), the value of $C_D A = 9.0$ ft^2. (If you are in a kind of fetal free-fall position then $C_D A = 2.5$ ft^2. If you are in a head-first, diving position then $C_D A = 1.2$ ft^2.)

Two questions. First, what is your free-fall terminal velocity, U_*? To answer this, we utilize equation (4.19) with $W = 200$, $\rho_a = 0.0024$, and $C_D A = 9.0$ to obtain $U_* = 136$ ft/s = 93 mi/hr.

Second, after how many seconds in the free-fall position do you attain terminal velocity? To answer this and related questions, we employ equations (4.13) and (4.14), noting that $U_* = 136$ ft/s and $y_0 = U_*^2/2g = 287$ ft.

It is easy to determine from equation (4.14) that after $t = 6$ s your fall velocity is $U = 121$ ft/s, which corresponds to $U/U_* = 0.89$. Your acceleration is $a = 6.7$ ft/s^2. From equation (4.13) we calculate that during this time you have fallen a distance $y = 450$ ft.

After $t = 11$ s, your velocity is $U = 134.5$ ft/s, your acceleration is $a = 0.7$ ft/s^2, and your fall distance is $y = 1{,}100$ ft. The velocity ratio, $U/U_* = 0.99$, is sufficiently close to the limiting value, $U/U_* = 1.0$, to conclude that $t = 11$ s is the answer to the question.

Wherein Dreadful Mistakes Are Made

Well, Joe Smog, a guy in our skydiving class, never gets anything right. On his first free-fall drop, he went into the head-first diving position instead of the required horizontal face-to-earth position.

PROBLEM 1. Assuming that $W = 200$ lb and $C_D A = 1.2$ ft^2 instead of 9.0 ft^2, show that Joe's terminal velocity was $U_* = 373$ ft/s (254 mi/hr), that he reached this velocity in about 30 seconds, and that, by that time, he had fallen over 8,000 feet.

Then Joe pulls the ripcord on his chute. Would you believe that the guy had put his parachute on upside down? This is almost impossible to do!

PROBLEM 2. In this case, the closed solid side of the hemisphere faced the air flow, not the hollow open side. So the drag coefficient was $C_D = 0.42$ instead of $C_D = 1.40$. (Again, this is not easy to do!) Show that Joe's terminal velocity was $U_* = 35.5$ ft / s (24 mi / hr), and that this is like jumping from a 20-foot wall when you hit the ground.

Fall of an Object from a Very High Altitude

Throughout the preceding analysis, we have assumed that the density of air, through which our object is falling, is constant. This assumption is valid as long as the object is not falling from too great a height. For example, at an elevation of, say, 500 meters, the density of air is about 95% of the sea-level density. However, at an elevation of 5,000 meters, it is only 50% of the sea-level value. Accordingly, for objects falling from very high altitudes, it can no longer be assumed that the air density is constant and so the preceding analysis is no longer valid.

An interesting study of the variable-density problem is given by Mohazzabi and Shea (1996). Their study of the phenomenon was motivated by the need to evaluate parachute performance in drops from 30,000 meters or so. As might be expected, the results they obtain are considerably different from those of the constant-density case.

5

Big Things Falling from the Sky

Within the category of big things falling from the sky, we have those phenomenal objects we call asteroids, comets, and meteors. And we all know that things descending from the heavens do not get much bigger than these.

In the introduction of chapter 1, we looked briefly at the incredible impact of a very high-velocity meteor in Arizona about 25,000 years ago. We shall examine this awesome event in more detail later in this chapter. In the meantime, let's carry out some computations involving objects somewhat less spectacular than meteors.

Golf Balls and Baseballs

Right now, our main task is to determine the kinetic energies of moving objects, regardless of whether they are raindrops or meteors. To accomplish this, we need to know the mass of an object and its velocity. Let us start with something simple, like golf balls and baseballs.

Golf Balls

The diameter of a golf ball is D = 1.65 inches and its weight is W = 1.62 ounces. In the centimeter-gram-second (CGS) system of units, these quantities are $D = 4.20$ cm and $W = mg = (46 \text{ g})$ $(982 \text{ cm/s}^2) = 45{,}000$ dynes. With the same analysis as that of chapter 4, we compute the terminal velocity of fall, U_*, of a golf ball from Newton's equation. The result is

$$U_* = \sqrt{\frac{8W}{\pi \rho_a C_D D^2}}, \tag{5.1}$$

where ρ_a is the density of air and C_D is the drag coefficient. For a smooth sphere, like a ping-pong ball, C_D depends only on the Reynolds number, $Re = \rho_a U_* D/\mu_a$, in which μ_a is the viscosity of air. This relationship between C_D and Re, for a smooth sphere, is shown in figure 4.1.

However, the dimples on a golf ball and the stitching on a baseball create a "roughness" that greatly affects the value of the drag coefficient. In another chapter, we shall look into this matter more closely in connection with trajectory analysis of golf balls and baseballs. For the present, the reasonable assumption is made that the drag coefficient of a golf ball is about $C_D = 0.25$.

Using $\rho_a = 1.24 \times 10^{-3}$ g/cm^3, it is easy to determine from equation (5.1) that the terminal velocity of a freely falling golf ball is $U_* = 4{,}580$ cm/s $= 45.8$ m/s. The kinetic energy is

$$\begin{aligned} e &= \tfrac{1}{2} m U_*^2 = \tfrac{1}{2}(46)(4{,}580)^2 = 48.2 \times 10^7 \text{ ergs} \\ &= 48.2 \text{ joules.} \end{aligned}$$

The tee-off velocity of a golf ball by a professional golfer can be as much as $U_0 = 70$ m/s. In this case, the kinetic energy of the ball is approximately $e = 112.7$ joules.

Baseballs

The diameter and weight of a baseball are $D = 2.90$ in and $W = 5.12$ oz. Converting these values to the CGS system, we obtain $D = 7.36$ cm and $W = mg = (145 \text{ g})$ $(982 \text{ cm/s}^2) = 142{,}000$ dynes. The roughness of the stitching on the baseball gives an approximate drag coefficient $C_D = 0.30$, over a broad range of Reynolds number. Again, from equation (5.1), we determine that the terminal velocity of a freely falling baseball has the value $U_* = 42.3$ m/s, and the kinetic energy is $e = 129.7$ joules.

A highly skilled professional pitcher is able to throw a baseball at a velocity $U_0 = 45$ m/s, and so $e = 146.8$ joules.

We observe that golf balls and baseballs have velocities of the order of 50 m/s and kinetic energies of the order of 100 joules. Perhaps you can relate to these numbers by imagining how much it would hurt if you got hit on the head.

We are interested in determining these velocities and kinetic energies of golf balls and baseballs because we want to compare them with the velocities and kinetic energies of other fast moving objects, such as hailstones, diving falcons, building smashers, charging elephants, cannon balls, automobiles colliding with walls, and, yes, large meteors!

Hailstones

Hailstones are usually created in those large beautiful white clouds called *cumulonimbus*, which frequently have heights exceeding 15,000 meters. These clouds are dangerous and pilots always try to avoid them. Normally there is much lightning and thunder activity in the very dark front of these anvil-shaped clouds.

Invariably, there are strong updraft currents in these formations. Indeed, it is this updraft feature, along with certain air temperature and moisture characteristics, that produces hailstones. A small drop of water, initially frozen into ice at a high elevation in the cloud, gradually grows in size as it falls with gravity and rises with updraft currents. Layers of water and entrained air are frozen into the ever-growing, falling-and-rising hailstone until it finally hits the ground.

Most everyone has seen hail and has been in a hailstorm. If the hail size is in the normal diameter range of 1.0 to 5.0 millimeters, a hailstorm can be almost enjoyable—unless you are driving along the expressway or you are a farmer. Great damage is done to crops each year by hail, especially in the states of the Midwest.

And, as fate would have it, the season of greatest hail activity —May through September—coincides with the most active period of crop growing and harvesting. It is estimated that hail and associated high winds cause about a billion dollars of damage to crops and property each year in the United States. Interesting descriptions of the structure and characteristics of hail and hailstorms are given by Flora (1956) and Knight and Knight (1971).

Hailstones are sometimes larger, considerably larger, than 5.0 millimeters diameter. There have been many reports of hailstones bigger than marbles, golf balls, and even baseballs. In this size range, they are definitely dangerous to livestock and even to people.

Probably the largest authenticated hailstone on record is the one that fell near Coffeyville, Kansas in September 1970. By great fortune, this hailstone was retrieved immediately and kept frozen until a silicone rubber mold could be made. This stone, labeled CK-1, weighted 766 grams and had a volume of 851 cubic centimeters. Its density $\rho_s = 766/851 = 0.90\ g/cm^3$, which is less than 1.0, reflects the presence of entrained air within its structure. Though not a perfect sphere (there were several bulges and bumps), the equivalent spherical diameter of CK-1 was $D = 11.75$ cm (4.62 in).

Over the years, plastic replicas of this giant hailstone, CK-1 have been studied by atmospheric scientists in various countries of the world. A research group in South Africa carried out wind tunnel experiments and free-fall drop tests from a helicopter. This study, reported by Roos (1972), indicates that the terminal velocity of CK-1 was approximately $U_* = 47.0$ m/s. A backward calculation utilizing equation (5.1) gives a drag coefficient $C_D = 0.52$, assuming sea-level air density; this seems about right.

Meteor Crater in Arizona

Without doubt, the most famous big thing ever to fall from the skies was the spectacular object that created, about twenty or twenty-five thousand years ago, what we now call Meteor Crater near the city of Flagstaff in northern Arizona.

Of course, at that time, there was no Flagstaff and there were few if any people around. Indeed, since the meteor collision occurred before the last ice age, there were no Great Lakes and no Saint Lawrence, Columbia, and Mackenzie Rivers. However, there certainly was that enormous ravine we call the Grand Canyon on the Colorado River about 150 kilometers to the northwest.

Still, as geologists measure things, 25,000 years is an extremely short period of time. Since the earth is more than four billion years old, this meteor impact event has taken place, you might say, during the final 0.02 seconds of an hour-long football game.

Now for some analyses and computations. The meteor was composed of iron with a small fraction of nickel. Accordingly, the density was $\rho_s = 7{,}850 \text{ kg/m}^3$. Assuming a spherical shape with diameter $D = 40$ m, the volume of the meteor was $V = (\pi/6)D^3 = 33{,}510 \text{ m}^3$. Consequently, the mass of the meteor was $m = \rho_s V = 263 \times 10^6$ kg = 263,000 metric tons. For comparison, the battleship *Iowa* displaces about 50,000 metric tons.

The meteor's velocity immediately prior to impact was $U = 72{,}000 \text{ km/hr} = 20{,}000 \text{ m/s}$. Therefore the kinetic energy was $e = (1/2)\ (263 \times 10^6)\ (20{,}000)^2 = 5.26 \times 10^{16}$ joules. This is about the same energy released from a thermonuclear bomb.

The several seconds just before and after the meteor's impact must have been absolutely awesome. A very interesting nontechnical description of the sequence of events is given by Grieve (1990).

When a large meteor strikes and enters the earth, intense shock waves are created in the meteor and the ground. The enormous kinetic energy of the meteor is instantly converted to pressures and temperatures so large that the meteor and adjacent zone are melted and even vaporized. Compression and expansion waves are transmitted downward and laterally to create a very large cavity in the ground. The cavity is quickly increased in size as rock and meteoritic material are ejected from the impact zone. Fragments of all sizes are projected upward and radially, the cavity attains its maximum depth and diameter, and a rim is established around the crater. Soon after, the crater wall partially collapses and much of the bottom is filled with fallen debris. In less than a minute, all is quiet.

The Meteor Crater of Arizona, seen in figure 5.1, has a diameter of about 1,250 meters and a rim height of approximately 50 meters above the surrounding plain. The present depth is about 170 meters, although the original depth was nearly double this amount.

Impressive as it is, Meteor Crater of Arizona is actually one of the smallest and newest of the impact craters studied by geologists and geophysicists. A list of about 120 terrestrial impact structures has been compiled by Grieve (1987), ranging in crater diameter up to 140 kilometers and in age to about two billion years.

No one knows how many large meteors have collided with the earth during the past four billion years. The traces of most have been erased by winds, rains, erosion, and sediment deposits.

FIG. 5.1

Meteor Crater, northern Arizona. (Photograph provided by Meteor Crater Enterprises, Inc.)

Simply on the basis of percentage areas, about 70% have plunged into the oceans. According to Grieve (1990), the only known marine impact crater is in the shallow water off the coast of Nova Scotia; it is fifty million years old and sixty kilometers wide. If you are interested in the probability and statistics of meteor impacts, Shoemaker (1983) will be very helpful.

Finally, one has only to look at some of the spectacular photographs of lunar landscapes to note the extensive history of meteor collisions on the moon. The absence of an atmosphere assures that craters stamped on the moon will be there forever. Scientists are able to gain much better understanding of meteoric activity on the earth by studying similar activity on the moon.

Interesting though rather technical articles concerning meteors and meteor craters are the presentations of Shoemaker (1983), Grieve (1987), and Melosh (1989).

Volcano Explosions

An entirely different kind of phenomenon, but one which nevertheless provides large numbers of "big things falling from the sky," is the explosion of a volcano. Maybe we can relate to these fantastic events more than we can to large meteor crashes because we can see big mountains and because volcanoes erupt rather frequently. In any event, we shall take a quick look at two very spectacular volcano explosions.

Mount Saint Helens

After several months of grumbling, Mount Saint Helens, a 2,950-meter volcano in southwest Washington State, finally woke up from its nap of about 125 years. On May 18, 1980, it suddenly exploded and sent 2.7 cubic kilometers of volcanic rock into the sky. More than 500 square kilometers of the surrounding region were devastated, an enormous avalanche was created, and rock fragments and volcanic ash were deposited hundreds of kilometers away. The top 400 meters were blown away, leaving a crater 750 meters deep and over two kilometers wide.

The total amount of kinetic energy released during the explosion of Mount Saint Helens on that fateful day was far greater than that of our meteor impacting near Flagstaff 25,000 years ago. Estimates of the kinetic energy of the Mount Saint Helens explosion range from about 1.0×10^{17} to 1.6×10^{18} joules. Perhaps the geometric mean value, $e = 4.0 \times 10^{17}$ joules, is representative.

Krakatoa

On August 27, 1883, a fifty-square-kilometer island called Krakatoa blew up and disappeared. Again, no meteors were involved. As in the case of Mount Saint Helens, there was an enormous volcanic explosion. Prior to the event, Krakatoa, situated in the very narrow Straits of Sunda between Java and Sumatra in what is now Indonesia, was a tropical island about two thousand meters high.

The explosion ejected about twenty-one cubic kilometers of rock, debris, and ash into the air. To give you a scale, this volume is equal to the volume of a cube with a side length of 2,750 meters (about 9,000 feet). Large chunks of rock of all sizes were blown into the atmosphere to heights of over twenty kilometers. Related shocks generated a huge ocean wave ("tsunami") nearly forty meters in height, which brought total destruction to nearby shores and took over 35,000 lives.

The total amount of kinetic energy released by the Krakatoa explosion was even more than Mount Saint Helens' energy. Estimated amounts range from 1.0×10^{18} to 6.2×10^{18} joules; the mean value is $e = 2.5 \times 10^{18}$ joules.

Incidentally, the kinetic energy of the Katmai volcano explosion in Alaska in 1912 was $e = 2 \times 10^{19}$ joules. Without doubt, the most spectacular of all in recorded history was the explosion of Santorini volcano in the Aegean Sea, north of the island of Crete, around 1600 B.C. The estimated kinetic energy was $e = 1 \times 10^{20}$ joules (about 1,200 thermonuclear bombs).

If you are interested in learning more about volcanoes you might want to refer to Decker and Decker (1981) for Mount

Saint Helens and to Simkin and Fiske (1983) for Krakatoa. For volcanoes in general and explosions in particular, Bullard (1976) is excellent.

Well, we have looked at a number of things falling from the sky, ranging in size from small to very large. Table 5.1 shows the amounts of kinetic energy associated with various events.

Meteor Crater Problems

The approximate cross section of Meteor Crater of Arizona is shown in figure 5.2. A reasonable assumption is that the profile is a portion of a circle. The radius of the crater is $a = 625$ m and the depth is $h = 170$ m.

PROBLEM 1. Utilizing the methods of integral calculus, show that the volume and area of the crater are

$$V = \frac{\pi h}{6}(3a^2 + h^2) \qquad \text{and} \qquad A = \pi(a^2 + h^2).$$

PROBLEM 2. Substitute the values $a = 625$ m and $h = 170$ m into these equations to obtain

$$V = 1.069 \times 10^8 \text{ m}^3 \qquad \text{and} \qquad A = 1.318 \times 10^6 \text{ m}^2.$$

PROBLEM 3. If the density of the ground material at Meteor Crater is $\rho_m = 2{,}350 \text{ kg/m}^3$, confirm that the mass of the material displaced by the meteor was about 250 million metric tons.

PROBLEM 4. We are considering the possibility of turning Meteor Crater into a football stadium. We shall use 15% of the total area calculated in problem 2 for the football field, stairs and aisles, rest rooms, and hot dog stands. We shall allow a 75-cm square (about 30 inches by 30 inches) for each seat. (This includes space for a small telescope so you can see the game.) Show that the crater can accommodate about 2,000,000 people. Obviously, this will be a fantastic place for the proposed Super Meteor Bowl game! And if you have been there, you know there is plenty of parking space for the 2,000,000 cars.

TABLE 5.1

Kinetic energies associated with many types of events

Event	$e = k \times 10^m$ joules	
	k	*m*
Raindrop (0.6 cm diameter)	5.0	−2
Hailstone (2.0 cm diameter)	1.0	0
Tennis ball serve by expert	6.0	1
High-speed golf drive or baseball pitch	1.3	2
Pelican diving for a fish	2.0	2
Football fast pass	2.0	2
Bowling-ball bowl	2.2	2
Giant hailstone (CK-1; 12 cm diameter)	8.5	2
Normal parachute landing	1.6	3
Professional football fullback plunge	3.2	3
Falcon diving at 200 mph	3.6	3
Pole vaulter landing on the pad	4.4	3
Cheetah running at 60 mph	1.6	4
Big steel ball smashing building	5.0	4
Pile driver for building foundation	6.0	4
Wall Street banker diving from Woolworth (1929)	1.6	5
Fast-charging big elephant	2.0	5
Old-time mortar cannon ball	5.0	5
Automobile collision into wall at 70 mph	9.0	5
B-25 collision with Empire State Building (1946)	1.0	8
16-inch projectile (impact only; no explosion)	3.5	8
16-inch projectile (1 ton TNT explosion)	4.2	9
Big ocean liner steaming at 20 knots	2.0	9
Towed iceberg (1,200 m long) moving at 2 knots	4.6	10

TABLE 5.1

(Continued)

Event	$e = k \times 10^m$ *joules*	
	k	*m*
Average-size tornado	1.0	11
Meteor impact in Arizona (25,000 B.C.)	5.3	16
Hydrogen bomb (20 megatons of TNT)	8.4	16
Average-size hurricane	1.0	17
Mt. St. Helens explosion (1980)	4.0	17
Krakatoa volcano explosion (1883)	2.5	18

Note: $e = (1/2)mU^2$.

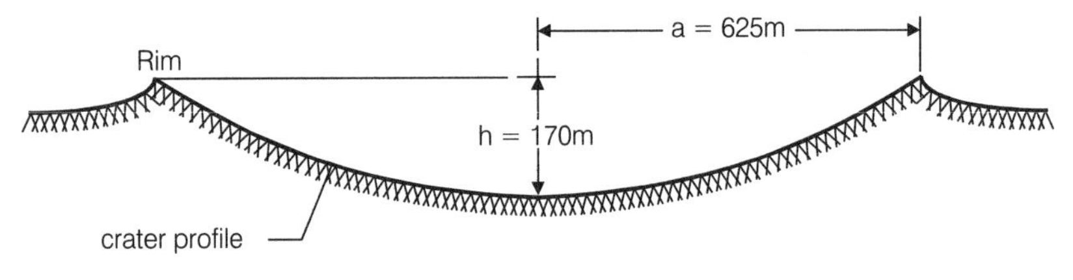

FIG. 5.2

Cross section of the Meteor Crater of Arizona

Big Things Falling from the Sky: A Scary Epilogue

On May 15, 1996, two University of Florida graduate students in astronomy discovered a large meteor, headed in the direction of the earth, moving at a velocity of about 58,000 kilometers per hour. They immediately reported their observations to the Harvard-Smithsonian Center for Astrophysics. After quickly confirming the location and projected trajectory of the meteor, the Center posted the relevant information on the World Wide Web. The meteor—technically speaking, it is called an asteroid—was named 1996 AJ1.

Sure enough, at 4:34 P.M. (GMT) on May 19, the meteor reached its point of closest approach to the earth. The distance:

about 450,000 kilometers. Astronomers consider this to be a very near miss; the distance to the moon is only 400,000 kilometers. It is extremely rare for a large meteor to come this close to earth.

Well, in the event you need something more to worry about, you can imagine what would have happened if 1996 AJ1 had indeed struck the earth. The meteor was estimated to be about 400 to 500 meters in size and composed of rock or iron, or, possibly, ice. We shall assume that the diameter of the meteor was $D = 450$ m and that its density (rock) was $\rho = 2{,}650$ kg/m^3. Accordingly, the mass of 1996 AJ1 was $m = \rho V = (\pi\rho/6)D^3 = 126.4 \times 10^9$ kg, where V is the volume and D is the diameter of the meteor. This computed mass is nearly 500 times larger than the mass of the meteor that crashed in Arizona 25,000 years ago.

The velocity of the meteor at its point of closest approach was $U = 16{,}100$ m/s. So the kinetic energy of the meteor on impact would have been $e = \frac{1}{2}mU^2 = 1.64 \times 10^{19}$ joules. This is about three hundred times greater than the kinetic energy of the Arizona meteor. Also, as is noted in table 5.1, this amount of energy is the equivalent of two hundred hydrogen bombs.

If the May 19 meteor had collided with the earth, there is a good chance it would have landed in the ocean since more than 70% of the world is covered with water. Even so, there is little doubt that the impact would have created massive tsunami waves in the ocean, which could have been very destructive.

Had the meteor hit a large city—for example, New York or Washington, D.C.—there would have been complete devastation. Utilizing scaling models presented by Shoemaker (1983), one can calculate that the crater produced by the May 19 meteor would have had a diameter $D = 8{,}500$ m, and a depth h = 1,200 m. The volume of the crater would have been about 34 billion meters^3. A direct hit on Washington, D.C. would have completely obliterated the entire central region of the city. The Potomac river would have quickly filled the crater to produce a large deep lagoon.

Fortunately, the meteor missed us. However, our worries may not be over forever. Since 1996 AJ1 has a four-year orbit it will be back in year 2000. Who knows? Maybe next time it will decide to come in for a landing.

6

Towing and Melting Enormous Icebergs: Part I

Where In the World Is the Ice?

The next topic we are going to examine involves the towing of icebergs. Now even though this rather intriguing method of obtaining fresh water has been talked about for quite a long time, nothing has ever been carried out with respect to actual engineering projects. However, it may well be that as the world gets much more crowded and people get much more thirsty, this source of fresh water will begin to look considerably more attractive. Then we may see some progress.

In any event, at the start of our iceberg towing studies we need to determine where in the world one finds a large amount of ice and a large number of icebergs. That's easy. Something like 90% of the annual production of ice in the world occurs on the ice shelves of the continent of Antarctica at the South Pole. The remaining 10% is produced in the Arctic region and the land masses surrounding the North Pole. More than half of this amount comes from the enormous island of Greenland—the world's second largest producer of icebergs.

Glacier discharges into the Atlantic along the shores of eastern Greenland produce icebergs that tend to move northward and are relatively small and few in number. However, glacier flows into the sea along the western coast of Greenland sometimes yield fifteen thousand icebergs in a single year, and a great many are very large. Most of these icebergs are discharged into Baffin Bay and are caught in the southward-moving Labrador Current, pass through the Davis Strait into the Labrador Sea, and finally emerge into the North Atlantic. As many as one thousand of them manage to reach the open ocean east of Newfoundland.

These Greenland icebergs come in all shapes and sizes; only the biggest ones reach the Atlantic. One of the largest was in the path of a merchant ship convoy during World War II and caused much havoc and confusion though no major disaster. It was 1,400 meters long, 1,100 meters wide, and 20 meters high. Since only the top one-eighth or so of an iceberg is above water level, the overall height of this giant was over 150 meters. Arctic-source icebergs as high as 150 meters (overall height 1,200 meters) have been reported.

Most of the Greenland icebergs that get as far south as latitude 45°N are caught in the warm waters of the northeast-ward-moving Gulf Stream. Though practically all of them are quickly melted, over the years a number have almost managed to reach Ireland. Suprisingly, several have somehow eluded the Gulf Stream and made it nearly to Bermuda.

Finally, it has been said that the Arctic is an ice-bound ocean ringed by continents, and Antarctica is an icy continent ringed by oceans. A remarkable description, in photographs and words, of these two incredible regions of the world is presented in a beautiful book by Stonehouse (1990).

A Prologue: Story of the Titanic

Just before midnight on April 14, 1912, the 46,000-ton west-ward-bound *Titanic*, on her maiden voyage, struck a very large iceberg at a location 800 kilometers south of Cape Race in Newfoundland. The iceberg caused substantial damage to the forward right side of the ship and severe flooding began immediately. The ship sank less than three hours later; over 1,500 lives were lost. Fascinating accounts of the tragedy are given by Eaton and Haas (1987) and Lord (1955).

For more than seven decades, the *Titanic* lay on the bottom of the Atlantic. Over the years, various attempts to locate her met with failure. Then, in September 1985, the vessel was finally discovered, about 30 kilometers southwest of her presumed location, at a depth of 4,000 meters and broken into two parts. The story of the ship's discovery is covered in the interesting book by Ballard (1987).

Equally interesting is the charming little volume by Brown (1983) entitled *Voyage of the Iceberg*. This is the story about the other major participant in the disasterous collision. It describes the origin of the massive iceberg, in the summer of 1910 in Jakobshavn Fjord in western Greenland, and its journey through Baffin Bay and Davis Strait and into the open Atlantic. For one brief minute, the iceberg and the vessel held their fatal rendezvous at 41°46′N, 50°14′W. Afterward, the iceberg continued its drift to the south until it finally melted and disappeared in the Gulf Stream.

It Is Time to Bring in the Numbers

We are now ready to begin the computations we need for our project. The total amount of water in the world is about $1,387.5 \times 10^{15}$ m^3. Of this, approximately $1,350 \times 10^{15}$ m^3—97.3%—is in the oceans and so, of course, it is all salt water.

The remaining 37.5×10^{15} m^3 is fresh water. About 29.0×10^{15} m^3—77.3%—is locked up in ice caps and glaciers. Another 8.37×10^{15} m^3 (22.3%) is in underground aquifers. The remainder of the fresh water is in the world's lakes and rivers (0.33%) and the atmosphere (0.035%). Of the 29.0×10^{15} m^3 of fresh water frozen in the world's glaciers and ice caps, 26.0×10^{15} m^3 (90%) is in Antarctica. The rest is in Greenland (8%) and in the Arctic (2%).

Scientists are not certain whether the total amount of ice in the world is increasing or decreasing. Never mind. If the amount is nearly in equilibrium, the total annual *increase* of ice from precipitation (snow and rain) is about equal to the total annual *decrease* by melting. It is estimated that these annual amounts of precipitation and melting are approximately $Q = 3.0 \times 10^{12}$ m^3/yr.

Incidentally, on this basis, the average residence time of ice in the world's glaciers and ice caps is $T = V/Q = (29.0 \times 10^{15})/(3.0 \times 10^{12}) = 9{,}700$ yr. This means that, on the average, a nice little snowflake that fell in Antarctica about ten thousand years ago only recently found its way to the ocean—and melted.

Back to the main point. We note that something like 3.0 trillion cubic meters of *fresh* water are produced annually from the melting of ice—mostly in Antarctica—and flow into the *salty* oceans. This amount is about equal to the 3.3 trillion utilized each year by humans worldwide for all purposes. Well, why not capture some of the ice, let it melt, and then catch the fresh water *before* it flows into the ocean? Good idea. Let us look into it.

Antarctica Icebergs

As mentioned, 90% of the world's ice is in Antarctica, a continent substantially larger than the United States and Mexico combined. Each year, something like 2.5 trillion cubic meters of ice from Antarctica's glaciers are fed into the southern sectors of the Atlantic, Indian, and Pacific Oceans. As shown on the map of Antarctica in figure 6.1, a great deal of this ice forms enormous ice shelves along the coasts of the continent. The major ones are the Ross (180° longitude), the Ronne-Filchner (45°W), and the Shackleton-Amery (75°E) ice shelves.

The the ice shelves begin to "calve." Large sections fracture, break off, and begin to drift in directions decided by water currents and winds. Some of these flat-topped icebergs are unbelievably large. One was located by the USS *Glacier* in 1956 about 200 kilometers west of Scott Island in the South Pacific (longitude 180°, latitude 67°S). This iceberg had a length of 330 kilometers, a width of 100 kilometers and an area of 30,000 square kilometers—larger than the state of Maryland. Another Antarctica iceberg, somewhat smaller, was discovered in 1946 and

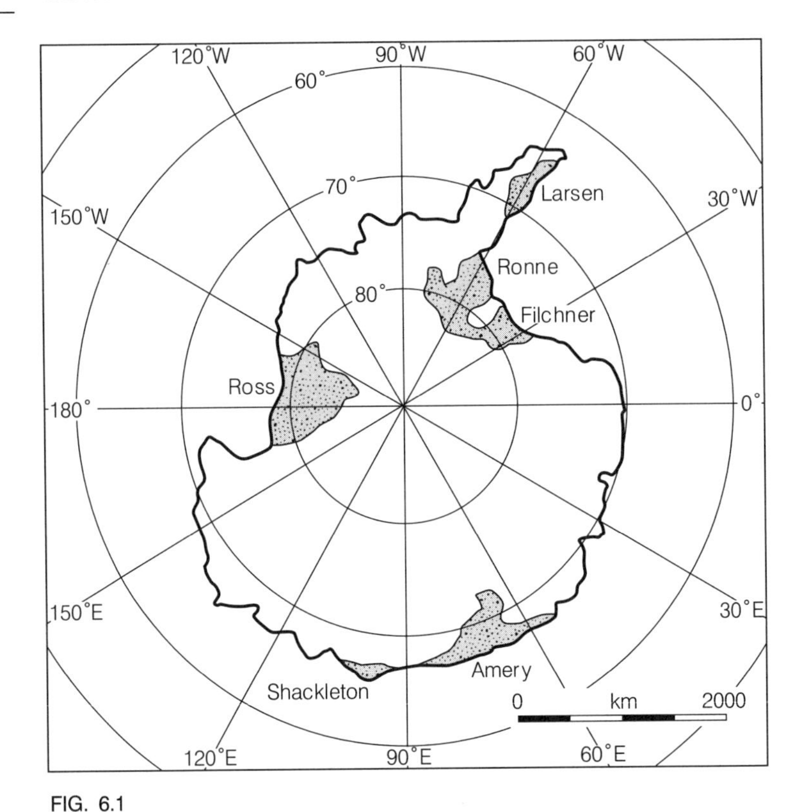

FIG. 6.1

Antarctica and its major ice shelves

was tracked for seventeen years. Nowadays, we can monitor movements of these giants from satellites. Needless to say, they are far too massive to tow.

Quite unlike the rugged Greenland icebergs, those of Antarctic origin are invariably "tabular" in shape, that is, they have flat horizontal top surfaces. A typical iceberg is shown in figure 6.2. These would be great as aircraft carriers or airports. Indeed, in the early days of World War II, the British government seriously considered using icebergs as "unsinkable aircraft carriers." During 1942–43, a 1:10 scale model was built and tested in Patricia Lake in Alberta, Canada.

Results of studies reported by Weeks and Mellor (1978) indicate that tabular icebergs of length about 1.0 kilometers and

FIG. 6.2

A tabular iceberg, 400 meters long, being moved by three U.S. icebreakers to clear a channel near the Ross ice shelf in Antarctica. (Photograph provided by U.S. Navy.)

width around 0.5 kilometers are quite common in the waters off Antarctica. A typical freeboard height (i.e., above water) is about 40 meters. The density of pure ice is about $\rho_i = 920$ kg/m^3. However, because of snow layers at the top of a tabular iceberg, the effective density can be as low as $\rho_i = 820$ kg/m^3.

An Introduction to Iceberg Towing

It is estimated that each year the world uses about 3.3 trillion cubic meters of fresh water. About two-thirds of this is for irrigation; the remainder is for municipal and domestic uses, manufacturing and mining, and electrical power plant water cooling. In many regions of the world, nature kindly provides ample, or more than ample, amounts of fresh water. However, in many other parts of our world, fresh water supply is far less than demand.

And where are the regions with great shortages of water? Clearly, all the desert and other arid zones, many of which are far from ocean shores. Some, though, are not very far: southwest Africa, Saudi Arabia, northern Chile and Peru, western Australia, and northwest Mexico. In addition, for example, there is that nearly parched, fast-growing place called southern California.

The map of Antarctica in figure 6.1 identifies the major ice shelves of the vast continent. An iceberg to be towed to the west coast of South America or Mexico or southern California would be acquired near the Ross ice shelf. For southwest Africa or the Middle East it would be obtained from the Larsen-Ronne-Filchner shelves, and for Australia we would go to Shackleton-Amery.

Wherein We Take Steps to Begin Commercial Operations

After lengthy periods of discussion and planning, we decide to proceed with our plan. We intend to capture large icebergs at the Ross ice shelf in Antarctica, tow them to Long Beach, California, let them melt, and then sell the nice fresh water to the thirsty people of Los Angeles, at a clear profit except for some expenses.

Well, as we shall see, there are a great many almost insurmountable engineering problems in the venture. Never mind; we proceed. Our next step is to organize our company, which we shall call Iceberg Towing and Water and Electricity Technology (IT WET).

Here is our plan after we establish the company. We talk the Navy into selling the USS *Iowa* and the USS *New Jersey* to Hertz, Ryder, Avis, or U-Haul. The cannons, turrets, and other heavy items are removed; this reduces the ship weights from 65,000 tons to about 50,000 tons. These ships are 270 meters long and 33 meters wide; maximum draft is 12 meters. Their top speed is around 35 knots and shaft horsepower 200,000.

Our new company IT WET will then lease these powerful ships from Hertz or whomever. They will be utilized to go down to Antarctica, catch icebergs near the Ross ice shelf, and tow them back to Long Beach. Here the icebergs will be melted and we will sell the fresh water to the Los Angeles water company. The schedules of the two ships will be arranged to assure that one iceberg is always being melted while another iceberg, being towed, will arrive just in time.

To follow almost the same numerical example as that of Weeks and Mellor (1978), the iceberg we select for towing will have the following dimensions, as shown in figure 6.3:

thickness, H: 200 m
freeboard, h_1: 25m
draft h_2: 175 m
width, B: 400 m
length, L: 1,200 m
volume, V: 96×10^6 m^3

The iceberg density is $\rho_i = 900$ kg/m^3 and the sea water density is $\rho_w = 1{,}205$ kg/m^3. The mass of the iceberg is $M_0 = \rho_i V = 86.4 \times 10^9$ kg. Ambient air and water temperatures are, say, 5°C.

Several fundamental questions are now asked. The first series of questions relates to matters involving the towing operation. The second series, in the next chapter, is concerned primarily with the melting process.

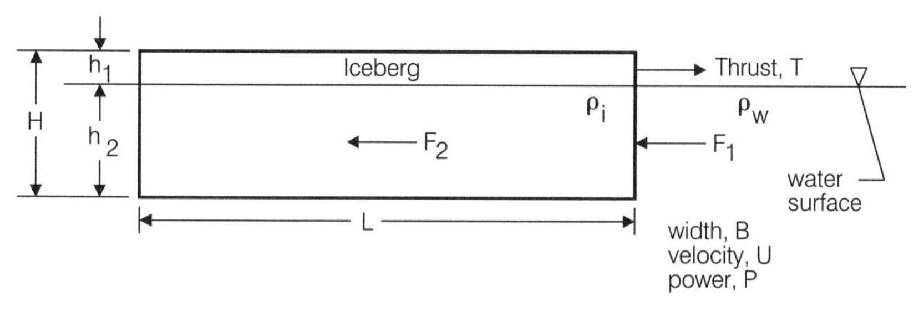

FIG. 6.3

Definition sketch for iceberg towing problem

Question 1

How far do we have to tow the iceberg? The coordinates of the origin (1), the Ross ice shelf, are longitude 150°W and latitude 75°S; and those of the destination (2), Long Beach, California, are longitude 118°W and latitude 34°N. To determine the distance between the origin and the destination, we use the following equation:

$$\cos a = \sin \phi_1 \sin \phi_2 + \cos \phi_1 \cos \phi_2 \cos(\lambda_1 - \lambda_2), \quad (6.1)$$

where a is the central angle, measured at the earth's center, of the great circle passing through the origin and the destination, and λ and ϕ are the respective longitudes and latitudes. Caution: be careful of the signs of the angles. The following sign rule is proposed: longitude, west $(+)$ and east $(-)$; latitude, north $(+)$ and south $(-)$.

Substituting the indicated values of λ and ϕ into equation (6.1) gives a 111°. Consequently, the distance from the Ross ice shelf origin to the Long Beach destination is $D = (111/360)2\pi(6{,}370) = 12{,}340$ km; $R = 6{,}370$ km is the radius of the earth.

Question 2

With what velocity should the iceberg be towed? There are two main problems regarding the velocity of towing. The first is: What power is required to tow it at velocity U, and does our ship have sufficient power to do this? We shall get back to this matter. The second is: We want to get the iceberg to Long Beach with minimum loss of ice due to melting. Is there a particular velocity that results in the least loss of ice from melting?

We construct an extremely simple mathematical model to examine this problem. With reference to figure 6.3, suppose that ice melting is caused by two main actions. The first is absorbed solar radiation by the nonwetted area of the iceberg (i.e., the top surface and freeboard sides). We assume that the melting loss due to radiation is directly proportional to the time of exposure, t_*. The second cause of melting is heat and mass transfer across the wetted area of the iceberg (i.e., the sides and bottom) due to the velocity of movement through the water. We assume these melting losses are proportional to U^2, as in many turbulent transfer processes.

For simplicity, we also assume that there are no changes in air and water temperatures during the journey. In a more realistic model, the effects of temperature change could be included.

So, with $t_* = D/U$, the following equation describes the relative loss of ice mass:

$$\frac{\Delta M}{M_0} = \frac{k_1 D}{U} + k_2 U^2, \tag{6.2}$$

where k_1 and k_2 are constants, t_* is the duration of the trip, and D is the distance. The quantity, $\Delta M = M_0 - M_d$ is the difference between the mass of ice at the origin and the mass at the destination. The first term of equation (6.2) describes the melting loss due to absorbed solar radiation on the top surface and freeboard sides; the second term is the loss due to turbulent convection along the submerged sides and bottom. In principle, the numerical values of the constants, k_1 and k_2, are known.

Look at equation (6.2). It says that if U is very small then the amount of melting due to absorbed radiation is large and that due to convection is small. The iceberg is essentially drifting; sooner or later it will melt entirely and disappear. On the other hand, if U is very large then the amount of melting caused by absorbed radiation is small but the amount due to turbulent convection is large. For example, if you want to melt the ice in your drink in a jiffy, stir it with vigor.

Then here is the question. Is there a velocity, U, for which the relative fraction melted, $\Delta M/M_0$, is a minimum? In other words, is there an "optimum" velocity, U_{opt}? This is the kind of question that arises in differential calculus. To obtain the answer we differentiate equation (6.2) with respect to U:

$$\frac{d(\Delta M/M_0)}{dU} = -\frac{k_1 D}{U^2} + 2k_2 U. \tag{6.3}$$

Setting this equation equal to zero gives an expression for U_{opt}, and substituting U_{opt} into equation (6.2) gives $(\Delta M/M_0)_{\min}$.

The answers are

$$U_{\text{opt}} = \left(\frac{Dk_1}{2k_2}\right)^{1/3},$$
$$\left(\frac{\Delta M}{M_0}\right)_{\min} = \left(\frac{27}{4}D^2k_1^2k_2\right)^{1/3}. \tag{6.4}$$

The distance of the iceberg tow is $D = 12.34 \times 10^6$ m. Also, we have $k_1 = 5.57 \times 10^{-9}\ \text{s}^{-1}$ and $k_2 = 0.0314\ \text{s}^2\ \text{m}^{-2}$. Substituting these numbers into equations (6.4) gives $U_{\text{opt}} = 1.03$ m/s = 2.0 nautical miles per hour = 2.0 knots and $(\Delta M/M_0) = 0.10 = 10\%$. Confession! These numerical values of k_1 and k_2 were computed in reverse, but only with the good intention of illustrating the point. The fact is, we do not know enough about iceberg melting to determine k_1 and k_2. If you would like to learn more about the complicated problems of water freezing and ice melting, the excellent book by Lock (1990) is recommended; it also contains a lengthy list of references.

Question 3

How long will it take to tow the iceberg from Ross ice shelf to Long Beach? The distance is $D = 12.34 \times 10^6$ m and the velocity is $U = 1.03$ m/s. So the towing time is $t_* = 11.98 \times 10^6$ s = 139 days or about 20 weeks. This is just right. For a twice-a-year iceberg delivery schedule by each of the two ships, we bring four icebergs annually to Long Beach. After each trip, each ship then has six weeks for repairs, replenishment of supplies, rest and recreation of crew, and the return journey to Antarctica. The

Iowa and *New Jersey*, at 30 knots, can get back down there in less than ten days.

Question 4

How much power will it take to tow the iceberg at a velocity $U = 2.0$ knots? After our ship has returned to the Ross ice shelf station in Antarctica, a suitable iceberg is located and attached securely to the towing cable. Then we apply towing power P to provide thrust T (see figure 6.3). The iceberg begins to move.

Newton's second law of motion says that

$$T - F = M_0 a = M_0 \frac{dU}{dt}, \tag{6.5}$$

where F is the resistance force, M_0 is the mass, and $a = dU/dt$ is the acceleration. It will take a period of time to accelerate the iceberg to full towing speed. However, we will assume that this has been done. Accordingly, $a = 0$ and so $T = F$.

For an object such as a ship or an iceberg moving on a water surface, the resistance force, F, is due to the two fluids air and water. For our iceberg problem we neglect air resistance.

Water resistance appears in three ways: (1) form drag, (2) skin friction, and (3) wave resistance. The latter can be neglected if the so-called Froude number, $Fr = U/\sqrt{2gH}$, is small, as it is in our iceberg problem. Form drag (F_1) and skin friction (F_2) can be described by equations of the form $F = \frac{1}{2}\rho_w C_* A U^2$, where C_* is a dimensionless coefficient and A is a reference area. For our problem we have

$$\begin{aligned} F &= F_1 + F_2 \\ &= \tfrac{1}{2}\rho_w C_D (B h_2) U^2 + \tfrac{1}{2}\rho_w C_F (2 L h_2 + BL) U^2, \end{aligned} \tag{6.6}$$

in which ρ_w is the density of sea water and C_D and C_F are, respectively, the drag coefficient and skin friction coefficient. The other quantities are defined in figure 6.3. The first term on the right-hand side of this equation describes the form drag due to the submerged area of the front of the iceberg; the second and third terms give the skin friction along the two sides and bottom of the iceberg.

For our problem we have the following numerical values: $\rho_i = 900\ \text{kg/m}^3$; $\rho_w = 1{,}030\ \text{kg/m}^3$; $H = 200$ m; $h_1 = 25$ m; $h_2 = 175$ m; $B = 400$ m; $L = 1{,}200$ m; $C_D = 1.20$; $C_F = 0.01$. The reference by Hoerner (1965) provides detailed information on form drag and skin friction coefficients. Substituting these numbers into equation (6.6) gives

$$F = (C_1 + C_2)U^2 = CU^2, \tag{6.7}$$

in which $C_1 = 43.3 \times 10^6$, $C_2 = 4.6 \times 10^6$, and $C = 47.9 \times 10^6$; F is the resistance force in newtons (N). We note that about 90% of the total resistance is due to form drag and 10% to skin friction.

After the iceberg acquires its constant towing velocity, the acceleration is zero, and, from equation (6.5), the thrust T is equal to the resistance force F. Accordingly, with $U = 2.0$ knots $= 1.03$ m/s, we have $T = F = 50.8 \times 10^6$ N.

The effective power required to tow the iceberg is $P_e = TU = (47.9 \times 10^6)U^3$. This equation says that when the velocity is doubled, for example, the power is increased eight times. Since $U = 1.03$ m/s, the *effective* power is $P_e = 52.3 \times 10^6$ N m/s $=$ 52,300 kW. (Note: newton meters/second = joules/second = watts.) The *shaft power* is $P_s = P_e/\eta$, where η is the propulsive efficiency of the ship's propellers. Since the *Iowa* class battleships were not designed to tow icebergs, we need to use a low value, say $\eta = 0.50$. Accordingly, $P_s = 105{,}000$ kW $= 140{,}000$ hp. The related shaft horsepower of the *Iowa* is about 212,000 hp so we can probably maintain a towing velocity of 2.0 knots. This assumes we have no problems with the iceberg, such as it breaking in two.

Question 5

What is the diameter, d, of the steel cable to connect the iceberg to the ship? This is one of a great many difficult problems of engineering in our iceberg towing venture. The thrust force in the cable is $T = 50.8 \times 10^6$ N. Assuming that the steel cable can tolerate a tension stress $\sigma = 35{,}000\ \text{lb/in}^2 = 2.41 \times 10^8\ \text{N/m}^2$, the cross-sectional area of the cable is $A = T/\sigma = 0.210\ \text{m}^2$, and so $d = 0.52$ m (20 inches). For comparison, each of the four main cables of the Brooklyn Bridge has a diameter of 16 inches.

Question 6

Pardon me, sir. Which way is Long Beach? Our big iceberg is securely attached to the towing cable which is securely attached to the *Iowa* or the *New Jersey*. We are now ready to leave the Ross ice shelf. Our ship will take the shortest route to Long Beach. This is the great-circle route; the distance is 12,340 kilometers. What should be the "heading" of our ship as it leaves Ross? We examine this important question, and numerous others, in the following chapter.

7

Towing and Melting Enormous Icebergs: Part II

Continuing our problem of towing an iceberg from Antarctica, we recall that the mighty *Iowa* had securely attached a long cable to an enormous iceberg and is ready to leave the Ross ice shelf (longitude 150°W, latitude 75°S) for its 12,340-kilometer journey to Long Beach (longitude 118°W, latitude 34°N).

The 50,000-ton *Iowa* with its 200,000 shaft horsepower, has already accelerated to the $U = 2.0$ knots constant towing speed, pulling the 1,700 times more massive iceberg behind it. The captain of the ship now intends to follow the great-circle route since this is the shortest distance to Long Beach. Questions: (1) what is the heading of the ship as it leaves Ross and (2) what is its heading as it arrives in Long Beach? By "heading" we mean the angle between true north (not magnetic north) and the course of the ship.

Drawing Triangles on Basketballs or Grapefruits

To answer these questions it is necessary to use spherical trigonometry. A definition sketch is presented in figure 7.1. This is a view of the world showing a spherical triangle with corners at the North Pole (N), Ross ice shelf (R), and Long Beach (L). The distances a, b, and c correspond to central angles, measured at

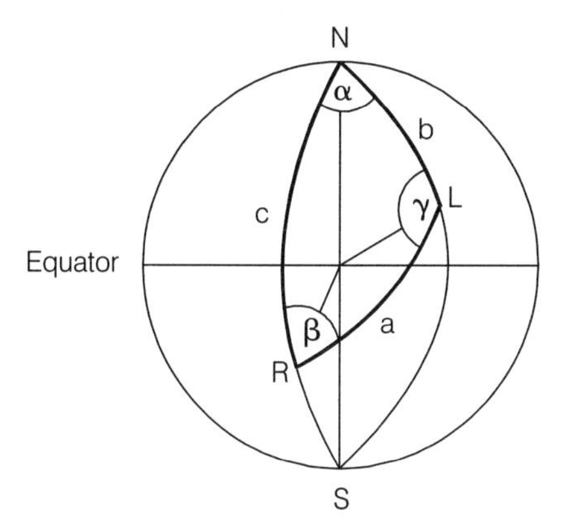

FIG. 7.1

Definition sketch of a spherical triangle

the earth's center, of arcs of great circles. The angles α, β, and γ are the angles between arcs *bc*, *ca*, and *ab*.

Take a look at this geometry on your globe of the world. Even better, lay out the spherical triangle on your basketball with string and sticky tape. Alternatively, use a grapefruit or cantaloupe and a marker pen.

Most problems of spherical trigonometry can be solved by using the law of cosines and the law of sines. The law of cosines is given by the relationships

$$\begin{aligned} \cos a &= \cos b \cos c + \sin b \sin c \cos \alpha, \\ \cos b &= \cos c \cos a + \sin c \sin a \cos \beta, \\ \cos c &= \cos a \cos b + \sin a \sin b \cos \gamma, \end{aligned} \tag{7.1}$$

and the law of sines is expressed by the equations

$$\frac{\sin a}{\sin \alpha} = \frac{\sin b}{\sin \beta} = \frac{\sin c}{\sin \gamma}. \tag{7.2}$$

For example, in our problem we know the following: (1) $a = 111°$; this was calculated in the previous chapter using the coordinates of Ross and Long Beach. (2) $b = 90° - 34° = 56°$, where 34°N is the latitude of Long Beach. (3) $c = 90° + 75° = 165°$ where 75°S is the latitude of Ross. (4) $\alpha = 150° - 118° = 32°$, where 150° and 118° are, respectively, the longitudes of Ross and

Long Beach. Substituting these numbers into the second of equations (7.1) gives $\beta = 28.2°$, and substituting into the third gives $\gamma = 171.5°$. So, looking at your globe or your basketball or grapefruit, the heading of the ship-iceberg as it leaves Ross is N28.2°E and the heading as it arrives in Long Beach is (180 – 171.5) = N8.5°E.

PROBLEM 1. With equations (7.1) and (7.2), compute the spherical coordinates (λ, ϕ) of the entire journey. Suggestion: let the angular distance a (the ship-iceberg path) change in small Intervals from 111° to 0°. Then, for each interval, calculate c (latitude), α (longitude), and β (heading). Keep in mind that $b = 56°$ and $\gamma = 171.5°$ remain constant.

The outcome of problem 1 is shown in figure 7.2; the solid line is the path of the ship-iceberg from Ross (R) to Long Beach (L). It may be a bit surprising to see a curved path since the ship is supposed to be taking the shortest journey: the great circle. The reason for this is that the map of figure 7.2 is the Mercator

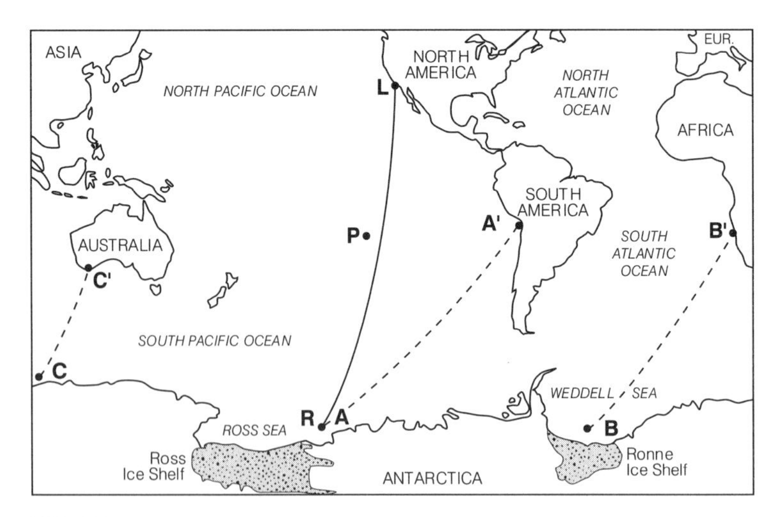

FIG. 7.2

Route of the iceberg tow from Ross ice shelf (R) to Long Beach (L). Pitcairn island (P) is approximately halfway. Shown also on this Mercator projection are three other possible iceberg tow routes.

projection of the world. In this case, as we would learn from study of a subject called cartography, it is impossible to transform a spherical surface onto a plane surface without some kind of distortion. The Mercator projection transforms the *variable* heading of a great circle on a sphere to a curved line on a plane. Furthermore, the Mercator transforms the *constant* heading of a path on a sphere to a straight line on a plane. This line is called a *rhumb line* or *loxodrome*.

This particular transformation property of the Mercator projection is an interesting and sometimes useful feature. However, the disadvantage of the Mercator is that areas in the polar regions become greatly distorted. For example, look at Antarctica in figure 7.2 and compare it with what you see on your globe or figure 6.1. The latter, by the way, is the so-called Lambert azimuthal equal-area projection.

It turns out that the route of our ship-iceberg comes close to the little island of Pitcairn (longitude 130°W, latitude 24°S). The area of the island is only five square kilometers and its 1995 population was fifty-five, mostly living in Adamstown, its capital. For such a tiny place it has quite a bit of history. You may remember that after the mutiny on the *Bounty* in 1790, the mutineers eventually made their way to and settled on Pitcairn. The present population, small as it is, is composed mostly of descendents of the original band of mutineers.

PROBLEM 2. Show that the great-circle distance from Ross (R) to Pitcairn (P) is $D_1 = 5{,}790$ km and from Pitcairn (P) to Long Beach (L) is $D_2 = 6{,}580$ km. Thus, the total distance, via Pitcairn, is 12,370 km, compared to 12,340 km direct.

PROBLEM 3. There would surely be a demand for the fresh water of Antarctic icebergs in places like northern Chile, southwest Africa, and western Australia. Select one of the following origin-destination schemes and carry out the necessary analysis.

A. Ross ice shelf (point *A* on the map of figure 7.2. 150°W, 75°S) to Iquique, Chile (point *A*′ on the map). This port city is adjacent to the vast Atacama desert, the driest place in the world. Assuming you can tow at 2.0 knots, how many days long is the journey?

B. Ronne ice shelf (point *B*, 45°W, 75°S) to Walvis Bay, Namibia (point *B*′). This destination would provide water to the Namib desert. At 2.0 knots, how many days?

C. Shackleton ice shelf (point *C*, 100°E, 65°S) to Esperance, Australia (point *C*′). This is about as close as you can get to the very arid zones of southwest Australia. Again, how many days if you tow at 2.0 knots?

The answers are (*A*) 85 days; (*B*) 75 days; (*C*) 42 days.

Spherical trigonometry is a valuable subject to learn if you are interested in cartography—the art and science of maps—or in surveying, navigation, or astronomy. If you want to understand space flight, astrodynamics, or celestial mechanics, it is absolutely essential to know the subject. The excellent book by Gellert et al. (1977) covers spherical trigonometry and a great many other topics of mathematics and its applications.

Interesting Topic: Coasting Icebergs

After 65 days of towing from the Ross ice shelf, the *Iowa* and its iceberg pass just east of Pitcairn island and after another 74 days, for a total of 139, the ship-iceberg approaches Long Beach. The *Iowa* disconnects the towing cable and heads for the nearby naval shipyard. The massive iceberg, now without the cable thrust force, nevertheless continues to coast toward the harbor. What *distance* will it travel before it comes to a stop?

The answer to this question is provided by Newton's equation. This relationship indicates that after the thrust force T is discontinued we have simply

$$-F = M_d U \frac{dU}{dx}, \tag{7.3}$$

in which $F = CU^2$ is the resistance force of the water, C is a constant, U is the instantaneous velocity, M_d is the mass of the iceberg, and $a = dU/dt = U\,dU/dx$ is the acceleration, where x is distance. With a bit of rearrangement, equation (7.3) becomes

$$\int_{U_0}^{U_s} \frac{dU}{U} = -\frac{C}{M_d} \int_0^{x_s} dx. \tag{7.4}$$

The lower limits of these integrals indicate that the iceberg velocity is $U = U_0$ when $x = 0$ (the location of the iceberg when the towing cable is released). The upper limits state that $U = U_s$ when $x = x_s$; the subscript s is explained below.

Integrating equation (7.4) and solving for x_s gives

$$x_s = \frac{M_d}{C} \log_e \frac{U_0}{U_s}. \tag{7.5}$$

From the information given in the previous chapter, we establish that the mass of the iceberg at destination is $M_d = 0.90M_0 = 77.8 \times 10^9$ kg. Also, assuming "linear shrinking" due to the 10% melting, the resistance coefficient is $C = 44.65 \times 10^6$. Theoretically the iceberg will never stop. However, for all practical purposes we can say its motion has ceased when its velocity is reduced to, say, 5% of the original velocity. Accordingly, we take $U_s = 0.05U_0$. Substituting these numbers into equation (7.5) gives $x_s = 5{,}200$ m or 5.2 km.

To carry the problem a step further, we change equation (7.3) to $-F = M_d\, dU/dt$. Then we set up the integral and solve it to obtain an expression for the time taken to stop. Taking $U_0 = 2.0$ knots $= 1.03$ m/s and $U_s = 0.05U_0$, we get the answer $t_s = 8.9$ hr.

How Do You Melt a Big Iceberg?

Capturing an iceberg of this enormous size and transporting it over twelve thousand kilometers through an entirely unpredictable and probably hostile ocean environment with no fracturing and minimum melting is an incredible marine engineering achievement. What do we do next with this gigantic piece of ice now coasting into Long Beach harbor? It is four times longer than the nearby *Queen Mary*, which is the second largest ocean liner ever built. The "flight deck" of our iceberg is thirty-five times larger than that of our biggest aircraft carrier. What do we do with this massive thing?

Well, we are faced with very difficult problems of melting the ice and delivering the fresh water. How is this done? How do we prevent the iceberg's melted fresh water from mixing with the ocean's salt water? Where do we get the heat needed to melt the iceberg? How do we effectively deliver the fresh water to the Los Angeles water company? And so on. Well, here are some schemes, that our company, IT WET, has considered.

Scheme 1

We air lift the big iceberg over to Meteor Crater near Flagstaff, Arizona. This will not be easy. Then we place the iceberg into the plastic-lined crater and just let it melt. By the way, this scheme sacrifices the "heat sink" value of the iceberg; we shall examine that topic shortly.

In any event, and with reference to problems 1 and 2 in chapter 5, to what depth will the now 10% melted iceberg fill enormous Meteor Crater? Answer: to about 145 m depth; the diameter of the circular lake will be 1,160 m in the 1,250-m-diameter crater.

Actually, this is not a bad scheme. They could certainly use the water in dry southern Arizona and we would have gravity flow all the way. Or we could pipe the water directly to the Colorado River about 150 kilometers to the northwest, again, by gravity flow.

Better still, we could simply put the iceberg into Lake Mead behind Hoover Dam and let it melt there. A word of caution: Our iceberg has about the same volume as Fort Peck Dam in Montana, America's largest earth dam. So it is necessary to put the iceberg into Lake Mead as gently as possible. In 1963 a very large land mass—about three times the volume of our iceberg—suddenly slid into the reservoir behind Vaiont Dam in northeast Italy. Although the very tall concrete arch dam (height 265 meters) was not destroyed, the landslide created a massive wave 100 meters high which went over the top of the dam and into the river valley below. More than three thousand lives were lost.

Scheme 2

We must be certain that the water in Long Beach harbor has a depth of at least 175 meters in the zone where the iceberg is to be anchored for harvesting. A curtain wall, to a depth of 25 meters or so, is placed around the boundary of the iceberg with a moat width of about 25 meters. Then the melting process is commenced using various thermal and mechanical devices to thaw the ice. The melted fresh water (say, 4°C, $\rho_f = 1{,}000\ \mathrm{kg/m^3}$) simply flows onto and remains on top of the slightly heavier ocean water (say, 15°C, $\rho_s = 1{,}030\ \mathrm{kg/m^3}$). The necessary pumping equipment is put into operation to extract and deliver the fresh water.

Again, this scheme sacrifices the "heat sink" value of the iceberg. Furthermore, it is not possible to safely utilize the warm sea water reliably as a heat source for melting. Most important, there is serious risk of fresh water contamination due to mixing at the fresh water–salt water interface, and by direct contact between the ice and warm ocean water.

Scheme 3

Even though the undertaking is very costly and time consuming, our company has decided to utilize the following scheme. We instruct our civil engineers to design and build two enormous "wet docks" the same size as our original iceberg ($200 \times 400 \times 1{,}200$ meters). By "wet dock" we mean a dock that is kept full, or nearly full, of water, like the ship locks in the Panama Canal. A "dry dock" is out of the question; to handle the entire hydrostatic force of 175 meters of water depth would require a structure the size of Hoover Dam. Indeed, as we shall see, a dry dock is not required for the operation.

After the *Iowa* disconnects and the iceberg coasts to a stop in the harbor, powerful tugs slowly move it into one of the docks (the other dock contains the nearly melted iceberg delivered by the *New Jersey* about three months earlier). The end gate of the dock is closed. The iceberg, now 10% melted, fits neatly into the dock. The melting operation is ready to commence. There are several important topics to consider.

Topic 1. The mass of the iceberg is $M_d = 77.8 \times 10^9$ kg. Our schedule requires that it be melted in three months. So the melting rate must be $m = 9{,}870$ kg/s, and hence the amount of fresh water produced is $Q = 9.87$ m^3/s.

The latent heat of fusion of water, that is, the amount of heat required to melt the ice, is 3.3×10^5 joules/kg. Consequently, the total amount of heat energy needed to melt the entire iceberg is $e = 2.6 \times 10^{16}$ joules $= 7.2 \times 10^9$ kWh (kilowatt hours). (For comparison, recall that the kinetic energy of the Arizona meteor on impact was around 5.3×10^{16} joules.) The amount of heat power required for melting is $p = 3.3 \times 10^9$ joules/s $= 3.3 \times 10^6$ kW. Our mechanical engineers will design the necessary heat transfer equipment to accomplish this task; the 15° to 20°C ocean water will be very useful for this melting process.

Topic 2. A promising technology for utilizing solar energy to generate power is the so-called ocean thermal energy conversion (OTEC) process. Briefly, this is a process that uses the relatively large difference in temperature between warm surface ocean water and cold deep ocean water. It works best in tropical regions like the Caribbean and the coasts of west Africa.

As Heizer (1978) indicates, this idea can be used in our ice melting operation. We have a turbine, for generating electrical power, driven by a working fluid such as ammonia or propane. Gaseous ammonia drives the turbine. Warm ocean water is used in the evaporator of the power generation cycle and—this is where the iceberg's heat sink comes in—the cold melted ice water is used in the condenser. So we have an ocean thermal energy conversion system. Utilizing this concept, we recognize that an iceberg has not one but two commodities of value: its fresh water and its "cold." Instead of simply allowing the cold melted ice to gradually warm up to ambient temperature, like an ice cube in a glass, we exploit its heat sink value.

We shall use this OTEC electrical power to operate machinery, pumps, a small stand-by desalination unit, and for other plant uses. For example, we need power to open and close the dock gates and to pump sea water through the heat exchangers. This OTEC power will be used also to pump the sea water, initially retained in the dock, back to the ocean as the melted cold water collects on top. This plant power is also available to operate a small desalination unit in case there is minor contamination by sea water. Thus, in IT WET, the E in WET refers to OTEC electricity.

Topic 3. The water melted from the iceberg is fed into a large fresh water tank and is ready for sale to the Los Angeles water company. A great many large and small chunks of ice, broken off the iceberg during the melting operation, are floating around in the fresh water tank. These will be collected and sold to local bars and restaurants at premium prices.

The design delivery of fresh water to Los Angeles is $Q = 9.87$ m^3/s. This corresponds to about 225 million gallons per day. Assuming a per capita consumption of 100 gallons per person per day, this amount will supply about 2.25 million people. The population of greater Los Angeles was about 14.5 million in 1995. So our iceberg project provides another 15% of the water needs of this very thirsty megalopolis.

Some Final Comments

All this discussion of iceberg towing started with the observation that nearly 80% of the world's fresh water is locked up in ice caps and glaciers, that each year an enormous amount of this ice melts and then mixes with the salty ocean water, and that it seems like a great idea to transport icebergs with their nice fresh water to places in the world where it is needed. If you are interested in the subject of iceberg towing, the book by Husseiny (1978) is recommended. It examines virtually every feature of iceberg utilization. A recent extensive review of numerous aspects of iceberg towing and melting is given by Wadhams (1996).

Staggering as the engineering problems are, it is safe to say that available or feasible technology makes iceberg towing *technically* possible. However, unless or until water shortages become extremely grave in certain regions of the world, it is likely that iceberg towing will not be *economically* competitive with desalination or long pipelines.

Finally, as interesting experiments, it would be worthwhile to tow a few icebergs of moderate size from, say, the Shackleton ice shelf to southwest Australia. This would provide valuable technical and economic information. However, in view of mankind's not always commendable past performance when dealing with nature, we should be absolutely certain that endeavors like massive annual amounts of iceberg transport from Antarctica or Greenland are, beyond any doubt, *ecologically* safe.

8

A Better Way to Score the Olympics

In response to the sports reporter's question, the coach replied, "Well, we don't know for sure, of course, but based on the results of our statistical analysis, there is a 90% probability that the team will set a new Olympics record, perhaps even a new world record." The sports reporter replied, "Well, good luck, coach."

Key words: for sure, of course, statistical analysis, about, probability, perhaps, luck

The subjects of probability and statistics are extremely important areas of the broad field of mathematics. In this chapter, and those that follow, we shall look at several topics which show how statistics and probability are used to analyze many kinds of phenomena and events. A complete list of the practical applications of statistics and probability would be endless: everything from the probability that it will rain tomorrow to your likelihood of winning at Las Vegas and from the annual cost of your life insurance to your chances of being kicked in the head by a horse or struck by lightning or attacked by killer bees.

From an almost infinitely long list of applications, we shall consider only a few. We start with an analysis of the medal scores of the 1992 Summer Olympic Games held in Barcelona. In subsequent chapters, we go on to fantastically interesting things like dropping a needle on a table to compute the numerical value of π, determining the probability that two people, within a certain size group of people, have the same birthday, calculating the minimum cost of having all your teeth extracted, counting the number of rice grains on a chessboard, and seeing how well a great many chimpanzees do behind a great many typewriters. But, for now, let's go to the Olympics!

We Need a Better Scorekeeper for the Olympics

In recent years we have observed that the Olympic Games have become increasingly nationalized, politicized, and commercialized. In addition, we have noted that preparation for and participation in the Games has become almost a whole new science. Wind tunnel studies are conducted to attempt to reduce the drag coefficients of bicycle riders and ski jumpers; mathematical models are devised to improve the biomechanics of high jumpers and pole vaulters; high-speed photography is employed to analyze the movements of gymnasts and relay racers; computer analyses are carried out to optimize the performance of kayak rowers and long-distance runners; and so on.

It seems as if everything relating to the Olympic Games is improving except for one thing: the system of final scoring of the participants. After all the incredibly hard work by the athletes and coaches and the countless hours of television viewing by billions of people around the world, all we get at the end is simply a dull column of numbers that tabulates how many medals each country has been awarded. A great many people believe that this denouement—this final outcome—is entirely inadequate.

We also read and hear a lot about the need for "level playing fields" in all kinds of arenas, especially economic and political. In no arena is this need greater than in the matter of determining the final scores of the Olympics. To illustrate this need, the following points and questions are raised:

1. The annual gross domestic products per capita (GDP/cap) of China, Nigeria, and Ghana are nearly identical (about $350). We

can say that the three countries are equally "poor." However, China has a population of 1,180 million, Nigeria 100 million, and Ghana 17 million. Thus, China has 70 times more people than Ghana from which to draw its athletes.

2. By the same token, the GDPs/cap of the United States, Canada, and Norway are about the same ($20,000). So they are equally "rich." But the population of the U.S. is 260 million, Canada 28 million, and Norway about 4 million. We note that the U.S. has a pool of athletes 65 times larger than Norway's.

3. Indonesia has a population of 195 million and GDP/cap of $700. Cuba has a population of 11 million and GDP/cap of $1,400. Qatar has a population of 0.50 million and GDP/cap of $17,000. Which country would be expected to receive the most Olympic medals: that country which is the poorest but most populous, that country which is the richest but least populous, or a country in between?

Values of the Medals

In the final tabulation of scores of nations competing in the Olympics, the numbers of gold, silver, bronze, and total medals are listed. The countries are then displayed in rank order according to the total number of medals. The results of the 1992 Summer Olympics are shown in table 8.1.

This system of ranking leaves much to be desired. For starters, the implication is that the three kinds of medals have the same "value." This is completely fallacious. Everyone knows that a gold medal is better than a silver medal is better than a bronze medal.

Accordingly, the following suggestions are made for a more realistic and rational assignment of values. We let w_1 = relative value of a gold medal, w_2 = relative value of a silver medal, and w_3 = relative value of a bronze medal.

1. Equal value. $w_1 = w_2 = w_3$. This is the present system.

2. Linear value. $w_1 = 3, w_2 = 2, w_3 = 1$.

TABLE 8.1

Final list of medals of the 1992 Summer Olympics

Nation	*Gold*	*Silver*	*Bronze*	*Total*	*Nation*	*Gold*	*Silver*	*Bronze*	*Total*
United Team	45	38	29	112	Ethiopia	1	0	2	3
United States	37	34	37	108	Latvia	0	2	1	3
Germany	33	21	28	82	Belgium	0	1	2	3
China	16	22	16	54	Croatia	0	1	2	3
Cuba	14	6	11	31	Iran	0	1	2	3
Hungary	11	12	7	30	Yugoslavia	0	1	2	3
South Korea	12	5	12	29	Greece	2	0	0	2
France	8	5	16	29	Ireland	1	1	0	2
Australia	7	9	11	27	Algeria	1	0	1	2
Spain	13	7	2	22	Estonia	1	0	1	2
Japan	3	8	11	22	Lithuania	1	0	1	2
Britain	5	3	12	20	Austria	0	2	0	2
Italy	6	5	8	19	Namibia	0	2	0	2
Poland	3	6	10	19	South Africa	0	2	0	2
Canada	6	5	7	18	Israel	0	1	1	2
Romania	4	6	8	18	Mongolia	0	0	2	2
Bulgaria	3	7	6	16	Slovenia	0	0	2	2
Netherlands	2	6	7	15	Switzerland	1	0	0	1
Sweden	1	7	4	12	Mexico	0	1	0	1
New Zealand	1	4	5	10	Peru	0	1	0	1
North Korea	4	0	5	9	Taiwan	0	1	0	1
Kenya	2	4	2	8	Argentina	0	0	1	1
Czechoslovakia	4	2	1	7	Bahamas	0	0	1	1
Norway	2	4	1	7	Colombia	0	0	1	1
Turkey	2	2	2	6	Ghana	0	0	1	1
Denmark	1	1	4	6	Malaysia	0	0	1	1
Indonesia	2	2	1	5	Pakistan	0	0	1	1
Finland	1	2	2	5	Philippines	0	0	1	1
Jamaica	0	3	1	4	Puerto Rico	0	0	1	1
Nigeria	0	3	1	4	Qatar	0	0	1	1
Brazil	2	1	0	3	Surinam	0	0	1	1
Morocco	1	1	1	3	Thailand	0	0	1	1

3. Quadratic value. $w_1 = 3^2 = 9$, $w_2 = 2^2 = 4$, $w_3 = 1^2 = 1$.
4. Cubic value. $w_1 = 3^3 = 27$, $w_2 = 2^3 = 8$, $w_1 = 1^3 = 1$.
5. Specific-gravity (s.g.) value. For gold s.g. is 19.30, for silver 10.50, and, for bronze 8.50. So $w_1 = 2.27$, $w_2 = 1.24$, $w_3 = 1$.
6. Intrinsic-worth value. For gold this is \$350/oz., for silver \$4/oz., and, for bronze \$0.10/oz. So, $w_1 = 3{,}500$, $w_2 = 400$, $w_3 = 1$.
7. Market-place value. A sportswear manufacturer or a fast-food chain will pay, to medal recipients, the following amounts for a television commercial: gold, \$100,000, silver, \$25,000, and bronze, \$1,000. So $w_1 = 100$, $w_2 = 25$, $w_3 = 1$.
8. Social value. A value system based on opinions of the athletes, their coaches, the various national Olympics committees, the International Olympic Committee, sportscasters, sports fans, and the general public.

The Proposed System of Scoring

First, to be conservative and simple, the *linear* value system is proposed: $w_1 = 3$, $w_2 = 2$, $w_3 = 1$. We shall use this value system in the following analysis.

Second, as suggested above, it is crucial that the *population* and the *economic status* of nations be taken into account. For example, Germany with a population of around 80 million and a GDP/cap of over \$17,000 would certainly be expected to receive far more medals than, say, Namibia with a population of 1.5 million and a GDP/cap of \$900. We need to level the playing field.

The final scores of the 1992 Summer Games are shown in table 8.1. If the scores are adjusted according to the linear value system (i.e., gold = 3, silver = 2, bronze = 1), the rank ordering of nations is not greatly altered. Some nations, which received a relatively large number of gold medals, advance in rank. These include Spain, Italy, Canada, and North Korea. In addition, ranking is better delineated for countries that had received a final score of only 1 or 2. For example, Greece, which acquired two gold medals, advanced from a score of 2 to a score of 6; Switzerland advanced from 1 to 3. In general, however, national ranking under the *linear* value system is about the same as under the present *equal* value system.

There is an entirely different outcome when population and GDP are taken into account. For the sake of brevity, the quite straightforward computation and tabulation are omitted. However, the essence of our analysis is the following. The adjusted score of each nation is divided by its population in millions. This gives a number, S_0. The gross domestic product per capita of a nation, GDP/cap, gives another number, G_0. These two numbers are plotted on so-called log-log paper, with G_0 as the abscissa (horizontal scale) and S_0 as the ordinate (vertical scale). The resulting plot is shown in figure 8.1. Each point identifies one of the sixty-four nations whose athletes received medals at the 1992 Games.

It is seen in the figure that the correlation between S_0 and G_0 is fairly weak, although we can detect a general trend of increasing S_0 with increasing G_0. Of course, the wide "scatter" of the data is a main point of our analysis. Some nations, large or small, rich or poor, receive high scores and others low scores. It is not a purpose of the analysis to determine *why* this is so.

A linear least-squares calculation of the sixty-four data points of figure 8.1 provides the equation

$$S_0 = 0.00182 G_0^{0.685}, \tag{8.1}$$

with a correlation coefficient of 0.5670. For the sake of a neater result, this expression is slightly modified to give

$$S_0 = 0.00215 G_0^{2/3}. \tag{8.2}$$

This equation, which we call the *expected score* relationship, is the desired result of our analysis.

What does this equation say? Well, take New Zealand as an example. The GDP/cap of New Zealand is \$9,710, that is, $G_0 = 9{,}710$. Substituting this number into equation (8.2) gives $S_0 = 0.979$ score points per million population. Since the popula-

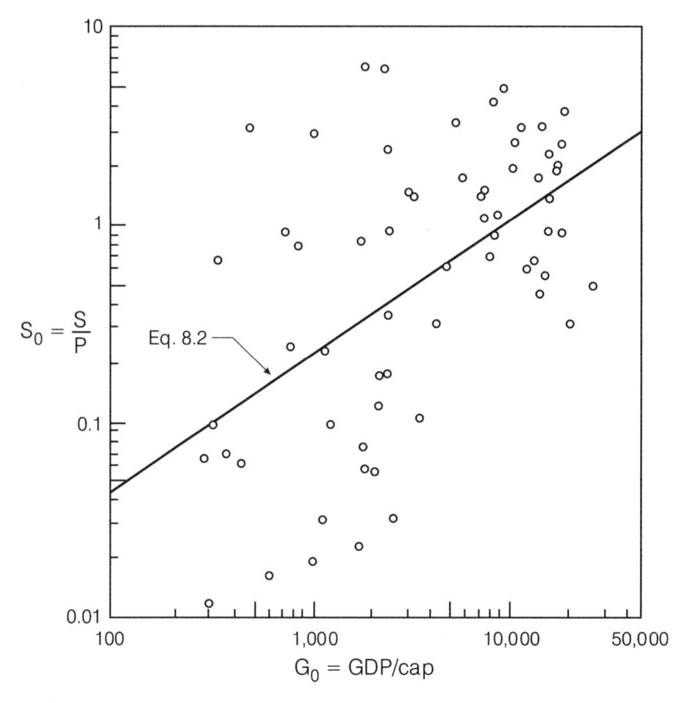

FIG. 8.1

Plot of the adjusted score per million people, S_0, versus gross domestic product per capita, G_0, for each country receiving medals at the 1992 Summer Olympics. The solid line is the expected score relationship.

tion of New Zealand is $P = 3.39$ million, then its *expected score* is $S' = (0.979)(3.39) = 3.3$. In fact, New Zealand had an *adjusted score* $S = 16$. Therefore, the *final score*, S_*, of New Zealand is, by definition, given by the expression $S_* = S - S' = 16 - 3.3 = +12.7$.

Similar calculations are carried out for the other sixty-three nations receiving medals. The outcome is that the scores of all nations are reduced and quite a few descend to negative values. Not unexpectedly, the final rank ordering of the nations is drastically altered. Looking at the original uncorrected scores of table 8.1, it may not come as a great surprise that Cuba advances to first place and Hungary to second. It turns out that Germany remains in third place, South Korea advances to fourth, and Australia to fifth. It is left as an exercise for the reader to determine the complete list of rank ordering.

Statistical Analysis of the Revised Scores

The preceding mathematics was mostly arithmetic. Now we get involved with a statistical analysis and we proceed as follows. First, we have before us the entire list of nations arranged in rank order of final scores, S_*, based on the above analysis.

Next, we decide to group the sixty-four nations into categories defined by a certain "class interval." For example, the high end of our score spectrum contains those countries with scores within the class interval +60 to +65. (There was only one: Cuba with +61.4.) Then we enter the countries with scores in the interval +55 to +60. (There were two: Hungary with +59.9 and Germany with +57.6.) And so on. We keep the 5.0 class interval width throughout. In some intervals there are no nations; in other intervals there are several or many.

Following this clustering exercise, we prepare a so-called histogram; this is shown in figure 8.2. The horizontal scale (abscissa) is the final score, S. (We drop the asterisk.) From right to left, this is the array of class intervals we just determined. The vertical scale (ordinate), p, of a particular rectangular block is equal to the number of countries within that block divided by 64 (the total number of countries). For example, in the interval $S = +60$ to $+65$, $p = 1/64 = 0.015625$; in the interval $S = +55$ to $+60$, $p = 2/64 = 0.03125$, and so on. We note that near $S = 0$, there are quite a few countries.

Well, we could stop here if we liked. The histogram tells us nearly everything we want to know: that only a few countries receive very high scores or very low scores; most of them receive average scores. But let us go a bit further. Precisely what do we mean by an "average" score and how are the scores spread out from high to low? To answer these questions, we turn to statistics.

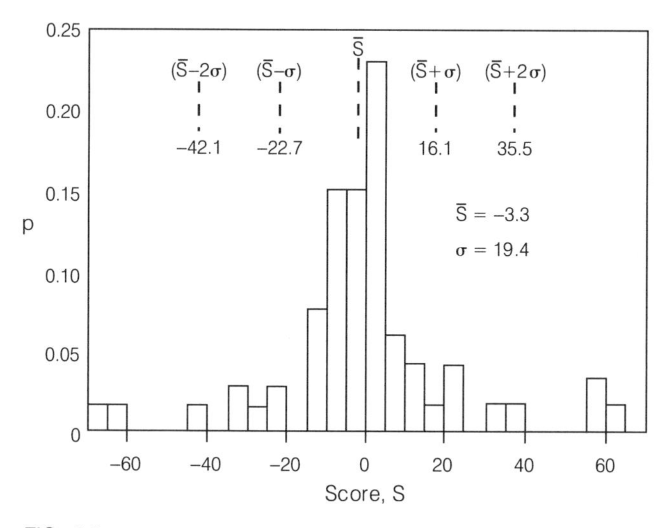

FIG. 8.2

Histogram of the final scores of the 1992 Summer Olympic Games

The average—or, better, *mean*—score can be computed from the equation

$$\bar{S} = \sum S_k p_k, \tag{8.3}$$

where the subscript k is the interval number. This expression says that the distance from the origin, $S = 0$, to the center of a particular class interval (plus or minus) is multiplied by the height of that class interval, p. For example, $S_1 = (62.5)(0.015625) = 0.9766$; $S_2 = (57.5)(0.03125) = 1.7969$. Then these products are added; the sum of all of them gives the mean value $\bar{S} = -3.359$. A statistician would say that this computation is the "first moment" about the origin. A structural engineer would call $\bar{S}$ the center of gravity.

Every student has suffered at the hands of wicked teachers who "grade on the curve." We shall be equally wicked in grading the Olympics. A few countries get an *A* or an *F*, several get a *B* or a *D*, most get a *C*. How do we decide? How do we spread out the grades in a logical and fair way—even though most teachers are entirely illogical and completely unfair?

The moment we say "spread out," the statistician immediately says: "Easy! compute the *variance* and the *standard deviation*." Okay, the so-called variance is calculated from the expression

$$\sigma^2 = \sum \left(S_k - \bar{S}\right)^2 p_k. \tag{8.4}$$

The standard deviation, σ, is the square root of the variance. Again, the statistician would say that this is the "second moment" about the mean and the structural engineer would call σ the moment of inertia. In any event, in our grading of scores we get $\sigma^2 = 375.40$ and so $\sigma = 19.375$.

Summarizing, the mean value, $\bar{S} = -3.3$, gives the average score and the standard deviation, $\sigma = 19.4$, provides a measure of how the scores are spread out. For example, we could say that all nations receiving scores within one standard deviation from the mean (i.e., from $S = +16.1$ to $S = -22.7$) get a C grade; those with scores between one and two standard deviations (i.e., from $S = +16.1$ to $+35.5$ and from $S = -22.7$ to -42.1) get either a B or a D; and those with scores larger than 2 get an A or an F. On this basis, for what it's worth, the final letter grades are A, 4 countries; B, 5; C, 48; D, 4; and F, 3 countries. These same numbers are provided by the histogram of figure 8.2.

Some Final Comments

We can put equation (8.2) into the following useful form:

$$S' = 0.215P^{1/3}G^{2/3}, \tag{8.5}$$

where S' is the expected score of a nation, G is its gross domestic product (GDP) in billions of U.S. dollars, and P is its population in millions of people. For example, the 1990 population of Greece was 10.05 million and its GDP was 48.05 billion dollars. Substituting these numbers in equation (8.5) gives $S' = 6.13$. That is, the expected score of Greece was $S' = 6.1$ points. The actual score of Greece, with its two gold medals, was $S = 6$. So, $S_* = S - S' = -0.1$. In other words, at the 1992 Olympic Games,

Greece's athletes performed exactly as would be expected. Nice tribute to the founder of the Games!

Finally, equation (8.5) is an interesting relationship for "leveling the playing fields." It says the following:

1. Two countries, 1 and 2, have the same population, that is, $P_1 = P_2$. However, country 1 has a GDP that is eight times greater than that of country 2, that is, $G_1 = 8G_2$. Accordingly, from equation (8.5), country 1 would be expected to receive a score four times larger than the score of country 2.
2. Two countries, 1 and 2, have the same gross domestic product, that is, $G_1 = G_2$. But the population of country 1 is eight times more than that of country 2, that is, $P_1 = 8P_2$. Accordingly, country 1 would be expected to receive a score twice as large as that of country 2.

Enough of the 1992 Summer Games. Now a word about where to go for more information. There are literally hundreds of books on statistics and probability. A great many are simple and easy to understand; others require at least a Ph.D. in mathematics to get beyond the first page. For a beginner, good places to start are the enjoyable books by Huff (1954), Moroney (1974), Mosteller (1965), and Weaver (1963). At the intermediate level, texts by Cooper and Weekes (1983) and Brandt (1976) are suggested. Somewhat more advanced are works by Dobson (1983) and Feller (1968). An excellent concise reference on statistical distributions has been prepared by Hastings and Peacock (1974).

PROBLEM. In February 1994, the Winter Olympic Games were held in Lillehammer, Norway. At these Games, athletes from twenty-two nations of the world received gold, silver, or bronze medals; the list of medals awarded by country is shown in table 8.2. As a homework assignment, analyze the results of the 1994 Winter Games in the same way as the preceding analysis of the 1992 Summer Games.

First, it will be necessary to convert the official results given in table 8.2 to the adjusted scores S, based on w_1 (gold) = 3, w_2 (silver) = 2, and w_3 (bronze) = 1. Next, you will need to look up the populations

TABLE 8.2

Final list of Medals of the 1994 Winter Olympics

Nation	*Gold*	*Silver*	*Bronze*	*Total*
Norway	10	11	5	26
Germany	9	7	8	24
Russia	11	8	4	23
Italy	7	5	8	20
United States	6	5	2	13
Canada	3	6	4	13
Switzerland	3	4	2	9
Austria	2	3	4	9
South Korea	4	1	1	6
Finland	0	1	5	6
Japan	1	2	2	5
France	0	1	5	6
Netherlands	0	1	3	4
Sweden	2	1	0	3
Kazakhstan	1	2	0	3
China	0	1	2	3
Slovenia	0	0	3	3
Ukraine	1	0	1	2
Belarus	0	2	0	2
Britain	0	0	2	2
Uzbekistan	1	0	0	1
Australia	0	0	1	1

and GDPs of the twenty-two nations to determine S_0 and G_0. A suggested reference is Wright (1996). For example, in table 8.2, it is easily determined that the adjusted score of Norway is $S = 57$. From our reference book, the population of Norway is $P = 4.297$;

TABLE 8.3

Final list of medals of the 1996 Summer Olympics

Nation	*Gold*	*Silver*	*Bronze*	*Total*	*Nation*	*Gold*	*Silver*	*Bronze*	*Total*
United States	44	32	85	101	Algeria	2	0	1	3
Germany	20	18	27	65	Ethiopia	2	0	1	3
Russia	26	21	16	63	Iran	1	1	1	3
China	16	22	12	50	Slovakia	1	1	1	3
Australia	9	9	23	41	Argentina	0	2	1	3
France	15	7	15	37	Austria	0	1	2	3
Italy	13	10	12	35	Armenia	1	1	0	2
South Korea	7	15	5	27	Croatia	1	1	0	2
Cuba	9	8	8	25	Portugal	1	0	1	2
Ukraine	9	2	12	23	Thailand	1	0	1	2
Canada	3	11	8	22	Namibia	0	2	0	2
Hungary	7	4	10	21	Slovenia	0	2	0	2
Romania	4	7	9	20	Malaysia	0	1	1	2
Netherlands	4	5	10	19	Moldova	0	1	1	2
Poland	7	5	5	17	Uzbekistan	0	1	1	2
Spain	5	6	6	17	Georgia	0	0	2	2
Bulgaria	3	7	5	15	Morocco	0	0	2	2
Brazil	3	3	9	15	Trinidad	0	0	2	2
Britain	1	8	6	15	Burundi	1	0	0	1
Belarus	1	6	8	15	Costa Rica	1	0	0	1
Japan	3	6	5	14	Ecuador	1	0	0	1
Czech. Rep.	4	3	4	11	Hong Kong	1	0	0	1
Kazakhstan	3	4	4	11	Syria	1	0	0	1
Greece	4	4	0	8	Azerbaijan	0	1	0	1
Sweden	2	4	2	8	Bahamas	0	1	0	1
Kenya	1	4	3	8	Latvia	0	1	0	1
Switzerland	4	3	0	7	Philippines	0	1	0	1
Norway	2	2	3	7	Taiwan	0	1	0	1
Denmark	4	1	1	6	Tonga	0	1	0	1
Turkey	4	1	1	6	Zambia	0	1	0	1
New Zealand	3	2	1	6	India	0	0	1	1
Belgium	2	2	2	6	Israel	0	0	1	1
Nigeria	2	1	3	6	Lithuania	0	0	1	1
Jamaica	1	3	2	6	Mexico	0	0	1	1
South Africa	3	1	1	5	Mongolia	0	0	1	1
North Korea	2	1	2	5	Mozambique	0	0	1	1
Ireland	3	0	1	4	Puerto Rico	0	0	1	1
Finland	1	2	1	4	Tunisia	0	0	1	1
Indonesia	1	1	2	4	Uganda	0	0	1	1
Yugoslavia	1	1	2	4					

consequently, $S_0 = S / P = 13.27$. The gross domestic product of Norway is \$76.1 billion. Accordingly, the gross domestic product per capita is $G_0 = \$17,700$.

Now you are ready to plot the (G_0, S_0) data points of the twenty-two nations. It is suggested you use log-log paper for this purpose.

Following this, you should try to determine the so-called line of best fit of these (G_0, S_0) data points by carrying out a least squares computation. This will give you a relationship for the expected scores, S_0, similar to equation (8.1) or equation (8.2) that we obtained for the 1992 Summer Games.

You can conclude your analysis by calculating the final score, $S_* = S - S'$, for each of the twenty-two countries. You should rank-order the nations according to their final scores and compute the average score $\bar{S}_*$ and standard deviation σ. Finally, you should prepare a histogram of the final-score distribution.

A Topic for a Term Paper

Perhaps you are taking a course in statistics, sports science, or some other subject and you are looking for a topic for the term paper you have to prepare. Well, here is a suggestion. How about carrying out a complete analysis of the scores of the Summer Olympic Games held in Atlanta in July 1996?

These Games were the largest and most comprehensive Olympics ever held; medals were given to athletes from seventy-nine nations. A list of the number of gold, silver, bronze, and total medals awarded is presented in table 8.3.

If you do decide to select this as the topic for your term paper, you should prepare a log-log plot of G_0 versus S_0, and determine the "best fit" equation relating these two quantities. You should also construct a histogram of the final scores of the seventy-nine nations. Finally, see if you can mathematically describe the histogram with the equation for the *normal probability distribution*. This very important distribution function is introduced and utilized in our next chapter.

9

How to Calculate the Economic Energy of a Nation

In this chapter, as we continue our study of statistics and probability, we shall examine a number of topics involving economics, physics, and geography.

Let's pick up where we left off in our previous chapter. Recall that we had generated an equation that allegedly predicts the score a nation should be expected to receive at the Olympic Games. By "score" we mean the summation of the weighted values of the gold, silver, and bronze medals a nation receives at the Games. The equation we got is

$$S = 0.215P^{1/3}G^{2/3}, \tag{9.1}$$

in which P is the population of a nation in millions of persons and G is its gross domestic product (GDP) in billions of U.S. dollars per year.

For example, in 1990 the population of Denmark was 5.14 million and its annual GDP was $94.7 billion. Substitution of these numbers in equation (9.1) gives $S = 7.7$. Thus, the expected score of Denmark at the 1992 Summer Games was 7.7. In fact, Denmark received a score of $S = 9$. Pretty close.

Now, let us give equation (9.1) a different interpretation. Let us say that it describes the relative *athletic strength* of a nation. If we adopt this point of view then, taking its population and GDP into consideration, the athletic strength of Denmark is about

$S = 8$. By the same token, with a 1990 population of 249.2 million and a GDP of $5,465.1 billion, equation (9.1) says that the athletic strength of the United States is $S = 420$. In the same way, $S = 201$ for Japan, $S = 111$ for Germany, $S = 52$ for Brazil, $S = 27$ for Mexico, and so on.

Fair enough. We now forget all about the Olympic Games and athletics and turn to economics. Dropping the constant, 0.215, in equation (9.1), we now say that the equation

$$S = P^{1/3}G^{2/3}, \tag{9.2}$$

provides a measure of the relative *economic strength* of a nation. Interesting. Where does this take us?

In successive steps, we change equation (9.2) into a form that may look familiar—at least to those who have studied a bit of physics. The steps are

$$\begin{aligned} S &= P^{1/3}G^{2/3} = (PG^2)^{1/3}; \\ S' &= S^3 = PG^2; \\ S'' &= \tfrac{1}{2}S' = \tfrac{1}{2}PG^2. \end{aligned} \tag{9.3}$$

We shall dispose of the symbols S' and S'' in a moment, so don't worry about them.

Kinetic Energy and Economic Energy

It is the last relationship of equation (9.3) that may ring a bell. In previous chapters we have examined various kinds of moving or falling objects, from baseballs to meteors, and, in numerous instances, we have calculated the kinetic energy of such objects. Recall that the equation for kinetic energy is

$$E = \tfrac{1}{2}mU^2, \tag{9.4}$$

where m is the mass of the object and U is its velocity. We note that this expression "looks like" the last relationship of equation (9.3). Let us write it in the form

$$E = \tfrac{1}{2}PG^2. \tag{9.5}$$

We shall call this quantity the *economic energy* of a nation.

The identical form of equations (9.4) and (9.5) establishes the basis for an analogy in which the mass m corresponds to the population P, and the velocity U corresponds to the annual gross domestic product G. A display of this analogy between the kinetic energy of a moving object and the economic energy of a nation is presented in table 9.1.

TABLE 9.1

Analogous quantities between the kinetic energy of a moving object and the economic energy of a nation

Quantity	*Physics: kinetic energy*	*Economics: economic energy*
Dimensions		
Force F	newton	keynes
Mass M	kilogram	person (millions)
Length L	meter	dollar (billions)
Time T	second	year
Force	Newton's second law	Keynes' second law
	$F = ma = m\dfrac{dU}{dt} = \dfrac{d(mU)}{dt}$	$F = Pa = P\dfrac{dG}{dt} = \dfrac{d(PG)}{dt}$
Units	newton $=$ kg m/s^2	keynes $=$ person dollar/yr^2
Momentum or impulse	$M = mU$	$M = PG$
Units	newton seconds	keynes years
Energy or work	$E = \frac{1}{2}mU^2$	$E = \frac{1}{2}PG^2$
	m: mass, kg	P: population, persons
	U: velocity, m/s	G: GDP, dollars/yr
Units	newton meters $=$ joule	keynes dollars $=$ veblen
Power	$P_* = \dfrac{dE}{dt}$	$P_* = \dfrac{dE}{dt}$
Units	joule per second $=$ watt	veblen per year $=$ ricardo

Two observations should be made here. The first is this. Our expression for economic energy, $E = (1/2)PG^2$, may or may not be an appropriate index of the economic status or capability of a nation. It seems reasonable, but why don't you think about it and decide? Make a fairly long list of countries, with their corresponding populations and annual GDPs and calculate their economic energies. Then start making comparisons. Perhaps the following examples will be helpful.

Example 1

The populations and the annual GDPs of Argentina and South Africa are about equal. So our equation, $E = (1/2)PG^2$, says that their economic energies are almost the same. Sound about right? Maybe so.

Example 2

The populations of the Netherlands and Saudi Arabia are nearly identical but the annual GDP of the Netherlands is nearly 2.5 times larger than that of Saudi Arabia. On this basis, the economic energy of the Netherlands is six times larger than Saudi Arabia's. Interesting. What do they have besides oil?

Example 3

The annual GDPs of Mexico and South Korea are roughly equal. However, Mexico has twice the population of South Korea. Therefore, from our equation, the economic energy of Mexico is twice as much as that of South Korea. Is this reasonable? Think about it.

Example 4

Here's a strange one. The population of New Zealand is 3.39 million and its annual GDP is $32.9 billion. The population of the Sudan is 25.20 million and its GDP is $11.4 billion. If you compute the economic energies of the two nations you will find that they are not greatly different. Does this make any sense? Excellent question. This example compares a relatively rich country with a small population and a relatively poor country with a large population. Our model says that they have about the same economic energy.

The second observation is this. In the display of table 9.1, the equivalent quantities between physics and economics are listed. The physics column is exact; the economics column consists of fabricated analogues.

The physics units—newton, joule, and watt—are real; they are named after noteworthy men of science and technology. The analogous economics "units"—keynes, veblen, and ricardo—are imaginary. Although they are named after famous economists, the units are invented here. It might be a good idea to request professional economists to deliberate and decide on appropriate names for the economics units.

In any event, this is a suitable place to cite a few references for those who would like to read about these notable economists, scientists, and engineers. *The Worldly Philosophers*, written by Heilbroner (1980), describes the lives, times, and ideas of a dozen famous economists, including Keynes, Veblen, and Ricardo. A book with the intriguing title *Flying Buttresses, Entropy and O-Rings* (Adams 1991) presents very interesting commentaries about Newton, Joule, and Watt and a great many other well-known scientists and engineers. Finally, the voluminous but easy-to-read *World Treasury of Physics, Astronomy and Mathematics*, edited by Ferris (1991), contains nearly one hundred chapters dealing with a wide range of scientific endeavors and the scientists involved in them.

Three Interesting Topics

Before we get back to statistics and probability, we look into three topics involving economic energy and economic power. The first topic is the following: Compare the economic energy of the twelve-nation European Community (EC) with that of the three-

TABLE 9.2

1990 populations and 1990 gross domestic products of the EC nations and the NAFTA nations

Nation	*Population P, millions*	*Gross domestic product (GDP), G, $ billions / yr*
European Community (EC)		
Belgium	9.85	143.1
Denmark	5.14	94.7
France	56.14	898.4
Germany	77.57	1,339.0
Greece	10.05	48.0
Ireland	3.72	27.4
Italy	57.06	765.5
Luxembourg	0.37	5.0
Netherlands	14.95	214.3
Portugal	10.29	37.5
Spain	39.19	301.8
United Kingdom	57.24	731.6
Total	341.57	4,606.3
North American Free Trade Agreement (NAFTA)		
Canada	26.5	440.1
Mexico	88.60	148.0
United States	249.22	5,465.1
Total	364.34	6,053.2

Source: Data from Wright (1996).

nation North American Free Trade Agreement (NAFTA). To make the comparison, the necessary information on populations and annual GDPs is listed in table 9.2.

Using our equation for economic energy, $E = (1/2)PG^2$, we determine that $E = 3.624$ billion veblens for the EC nations and $E = 6.675$ billion veblens for the NAFTA. Thus the NAFTA economic energy is nearly 85% larger.

The second topic we want to examine involves economic power—with the connotation that we are talking scientifically, not politically. In the list of analogous quantities in table 9.1, we note the entry: Power, $P_* = dE/dt$. As this equation indicates, power is the time rate of change of energy, that is, the *rate* at which energy is produced or utilized.

For example, determine the economic power of the United States, at five-year intervals, during the period 1960 to 1990. The necessary numbers and results of computations are shown in table 9.3. We observe, from the right-hand column of the table, that the economic power of the United States increased from

TABLE 9.3

Economic energy and economic power of the United States, 1960 to 1990

Year	*Population* P *million persons*	*GDP* G *billion dollars / yr*	*Energy* $E = \frac{1}{2}PG^2$ *million veblens*	*Δ Energy* ΔE *million veblens*	*Power* $P_* = \frac{\Delta E}{\Delta t}$ *million ricardos*
1960	179.3	515.3	23.8		
1965	190.9	705.1	47.5	23.7	4.7
1970	203.3	1,015.5	104.8	57.3	11.5
1975	214.6	1,598.4	274.1	169.3	33.9
1980	226.6	2,732.0	845.7	571.6	114.3
1985	237.8	4,014.9	1,916.6	1,070.9	214.2
1990	249.2	5,465.1	3,721.5	1,804.9	361.0

Source: Data from Wright (1996).

about 4.7 million ricardos during 1960–65 to around 361 million ricardos during 1985–90.

By the way, if you are interested in such things, the economic energy data (E) and the economic power data (P_*) of table 9.3 both plot as fairly nice exponential growth curves. You might want to confirm this. Also, when will this exponential growth begin to flatten out to logistic growth? Bound to happen, sooner or later.

The third topic to be analyzed involves comparisons of the economic energy of a nation and the kinetic energy of a moving object. First, from table 9.3, we note that the economic energy of the United States in 1990 was $E_{ea} = 3{,}722 \times 10^6$ veblens. The subscript *ea* refers to economic (e) and to the United States (a). Next, in the statistical analysis later in the chapter, 150 nations are examined. The nation with the smallest economic energy is the Maldives, a country composed of over a thousand very small islands (total area 300 square kilometers) in the Indian Ocean about 400 kilometers southwest of the southern tip of India. Its 1990 population was 222,000 and its annual GDP was $65 million. So, the 1990 economic energy of the Maldives was $E_{eb} = 469 \times 10^{-6}$ veblens. The ratio of the economic energies of these two extremes, the United States and the Maldives, is $E_{ea}/E_{eb} = 7.94 \times 10^{12}$ (about eight trillion).

Now, in chapter 5, we determined that the kinetic energy of the Arizona meteor, immediately prior to impact, was $E_{ka} = 5.26 \times 10^{16}$ joules. This time the subscript *ka* means kinetic (k) and meteor (a).

You probably see the question coming. Here it is. What would be the kinetic energy of a falling object that has the same ratio to the kinetic energy of the Arizona meteor as the ratio of the economic energies of the Maldives and the United States?

Well, from the above numbers we easily establish that the answer is $E_{kb} = 6.62 \times 10^3$ joules. We also establish that this is the kinetic energy of a fifty-pound weight falling from a ten-story building. Contrast this kinetic energy with that of the Arizona meteor and its mass of 263,000 metric tons moving at a velocity of

20 kilometers per second. Clearly, this is like comparing the economic energies of the Maldives and the United States.

Going one step further, it turns out that the mean economic energy of the 150 countries we shall examine shortly is about $E = 346$ veblens. This value is midway between Guatemala with $E = 281$ veblens and Sri Lanka with $E = 416$.

Staying with our earlier maximum values—the Arizona meteor (for kinetic energy) and the United States (for economic energy) —the value of $E = 346$ veblens corresponds to $E = 4.89 \times 10^9$ joules. From table 5.1 this is about the same kinetic energy as that released in an explosion of one ton of TNT. Alternatively, if you prefer, it is approximately the same kinetic energy as that of a 1200-ton locomotive falling from the top of Sears Tower in Chicago, which is quite unlikely.

A Generalized Equation for Economic Strength

We now set aside the topic of kinetic energy of moving bodies and return to equation (9.2),

$$S = P^{1/3}G^{2/3}.$$

You will remember that we called this the economic strength of a nation. In this form, this equation is simply an arithmetic modification of the equation for economic energy, E.

A slight diversion. Let us rewrite equation (9.2) as follows:

$$S = P^{1/n}G^{(n-1)/n}, \qquad (9.6)$$

where n is a number equal to or greater than 1.

This form provides a mechanism to argue about the merits of our economic strength or energy model. Do you think we placed too much emphasis on annual GDP and not enough on population? Then select and use $n = 1$ in equation (9.6). This gives simply $S = P$, which says that the economic strength of a country is numerically equal to the population and that the GDP is not important. That's silly.

So try $n = 2$. This yields the expression $S = (PG)^{1/2}$. With reference to table 9.1, this result indicates that economic strength is proportional to the square root of the *economic momentum* of a nation. Maybe so.

Next, try $n = 3$. Clearly, this takes us back to equation (9.2), which is what we are using here.

Do you want more emphasis on annual GDP, G, and less on population, P? Then take $n = 4$, 5, 6, 25, or 100. Indeed, if you select $n = \infty$, our equation becomes $S = G$. This answer indicates that economic strength depends only on annual GDP and that population does not matter. That's silly also.

Well, you select whatever value of n you like. We shall stay with $n = 3$ for the rest of our analysis.

Back to Histograms

Here is what we are going to do. First, we get a suitable reference book to provide recent information about the populations and annual gross domestic products of the nations of the world. We then list the nations by continents: Africa (51), Europe (26), South America (12), Asia (39), North America (17), and Oceania (5). The numbers in parentheses indicate the number of countries; the total is 150.

For each nation we tabulate the values of population, P, in millions of persons, and the GDP, G, in billions of U.S. dollars per year. For each we then compute $S = P^{1/3}G^{2/3}$.

Next, in order to shrink the extremely wide range of S, we calculate $S_* = \log_e S$. That is, we determine the natural logarithm of S and then consider that S_*, not S, is the parameter that describes economic strength. Essentially, this is what is done when intensities of earthquakes are reported on the Richter scale.

As we did in chapter 8, we then establish "class intervals." Our range of S_* is from $+7.50$ (U.S.) to -2.32 (Maldives). So we select class intervals -2.50 to -1.50, -1.50 to -0.50, -0.50 to $+0.50$, and so on, to the interval $+6.5$ to $+7.5$.

Now we get to the statistical analysis and we carry it out as we did in the previous chapter. The mean value, $\bar{S}_*$, is

$$\bar{S}_* = \frac{1}{N_0} \sum S_{*,k} N_k, \tag{9.7}$$

in which N_0 is the number of nations, $S_{*,k}$ is the distance between the origin $S_* = 0$ and the midpoint of a particular class interval (-2.0, -1.0, 0, $+1.0$, and so on), and N_k is the number of nations within that class interval.

The variance, σ_*^2, is calculated from the expression

$$\sigma_*^2 = \frac{1}{N_0} \sum \left(S_{*,k} - \bar{S}_*\right)^2 N_k. \tag{9.8}$$

The standard deviation is the square root of the variance. Another important statistical parameter is the *coefficient of variation*, which has the definition

$$\text{C.V.} = \sigma_* / \bar{S}_*. \tag{9.9}$$

This parameter provides a measure of how broad or how narrow is the spread of a particular statistical distribution in comparison with the mean value.

The results of our computations are summarized in table 9.4.

The next step in our analysis is to prepare histograms. We could construct seven histograms if we wished—the six "continents" plus the world—but let us settle for two: Africa and the world. These are shown in figure 9.1. For the moment look only at the "bars"; disregard the smooth curves. By the way, if you need something to do sometime, you can construct the other five histograms.

In both histograms you will note that there are very few countries with large values of S_* and very few with small values. Most of the nations are centered around the mean value, $\bar{S}_*$. Not surprisingly, the mean value for Africa is substantially less than the mean value for the world. This is evident from the numbers shown at the bottom of table 9.4.

TABLE 9.4

Results of a statistical analysis concerning the economic strengths of the nations of the world

Class interval S_*	*Africa*	*Asia*	*Europe*	*North America*	*South America*	*Oceania*	*World*
$-2.5/-1.5$	2	1		1		1	5
$-1.5/-0.5$	4	1			1		6
$-0.5/+0.5$	11	2		2	2		17
$+0.5/+1/5$	15	7	3	6		2	33
$+1.5/+2.5$	12	8		4	3		27
$+2.5/+3.5$	3	7	6	1	2	1	20
$+3.5/+4.5$	4	8	11		3	1	27
$+4.5/+5.5$		2	1	2			5
$+5.5/+6.5$		2	4		1		7
$+6.5/+7.5$		1	1	1			3
Total	51	39	26	17	12	5	150
$\bar{S}_*$	1.10	2.62	3.77	1.88	2.42	1.40	2.18
σ_*	1.43	1.89	1.51	2.08	1.94	2.06	1.98
C.V.	1.30	0.72	0.40	1.11	0.80	1.47	0.91

Equally evident from these numbers is that the standard deviation σ_* for Africa is the least for any of the continents. These results indicate that the economic strengths of virtually all of the African nations are well below those of the other continents. At the other extreme is Europe, which has the largest value of $\bar{S}_*$. With a standard deviation about the same as Africa's, it can be said that the economic strength of virtually all of the European nations are well above those of the other regions of the world.

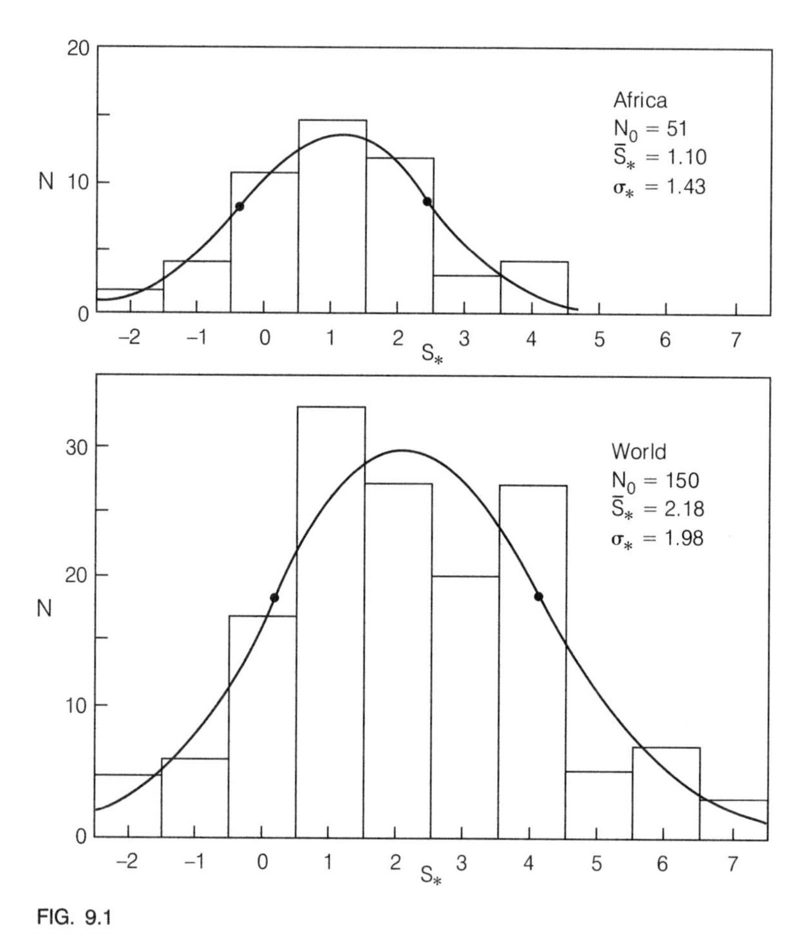

FIG. 9.1

Histograms of the economic strengths of the nations of Africa and the world. $S_* = \log_e S$ and $S = P^{1/3}G^{2/3}$.

The Probability Distribution Curve

We now carry out an exercise in "curve fitting." That is, can we describe the data displayed in the "bar chart" of the histograms by a continuous mathematical equation?

Well, looking at the two histograms, especially that for Africa, it appears that the *normal probability distribution* may do the job.

This distribution has the definition

$$N = \frac{N_0}{\sigma_* \sqrt{2\pi}} e^{-(S_* - \bar{S}_*)^2 / 2\sigma_*^2}, \tag{9.10}$$

where N_0 is the total number of nations (e.g., $N_0 = 51$ for Africa). We already know the values of $\bar{S}_*$ and σ_*; these are listed at the bottom of table 9.4. So we simply plot the curves $N = f(S_*)$. The results are the smooth curves shown in figure 9.1. The mathematical curves "fit" the computed data fairly well.

As we could easily show, one of the properties of the normal distribution is that the *inflection point* (i.e., the point where the curve changes from being concave downward to being concave upward) occurs at $S_{*,i} = \bar{S}_* \pm \sigma_*$, $N_i = N_0/\sigma_* \sqrt{2\pi e}$. These points are shown as the solid dots in figure 9.1. Again, the bell-shaped normal distribution appears to match the data.

A suggested reference that deals with the normal probability distribution curve is the little book by Rozanov (1977).

A Suggestion for a Term Paper

Now that we have examined the economic strengths of 150 nations of the world, diligent students may want to carry out a

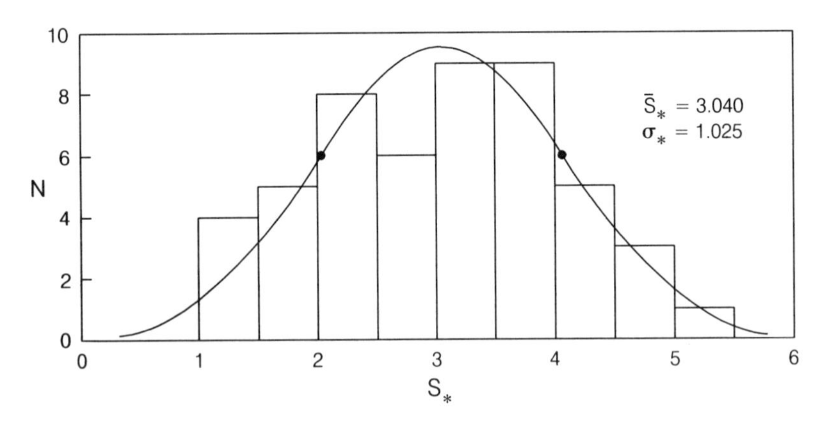

FIG. 9.2

Histogram of the economic strengths of the fifty states

TABLE 9.5

Economic strengths of the states of the United States

	State	*S*		*State*	*S*
1	California	230.91	27	South Carolina	22.06
2	New York	143.73	28	Oklahoma	20.02
3	Texas	116.05	29	Oregon	19.77
4	Florida	94.14	30	Iowa	18.85
5	Illinois	87.39	31	Kansas	17.46
6	Pennsylvania	85.57	32	Mississippi	14.59
7	Ohio	74.92	33	Arkansas	14.27
8	New Jersey	67.02	34	Nebraska	10.88
9	Michigan	65.96	35	West Virginia	10.55
10	Massachusetts	48.36	36	Utah	10.54
11	Virginia	46.27	37	New Mexico	9.35
12	Georgia	44.42	38	Nevada	9.20
13	North Carolina	43.93	39	Hawaii	8.72
14	Maryland	38.26	40	New Hampshire	8.40
15	Indiana	37.41	41	Maine	8.26
16	Washington	36.29	42	Rhode Island	7.11
17	Missouri	35.23	43	Idaho	6.43
18	Wisconsin	34.10	44	Montana	5.14
19	Tennessee	31.89	45	Delaware	5.07
20	Minnesota	31.69	46	South Dakota	4.54
21	Connecticut	28.79	47	Alaska	4.47
22	Louisiana	26.03	48	North Dakota	4.05
23	Alabama	25.49	49	Vermont	3.86
24	Colorado	24.41	50	Wyoming	3.06
25	Arizona	24.21		Entire U.S.	1,952.53
26	Kentucky	23.12			

Source: Data from U.S. Bureau of the Census (1994).

similar analysis of the 50 states of America. For a course in statistics, economics, or geography, this could make a very interesting report.

From an up-to-date information source, you will need to look up the population and the GSP (gross state product) of each state. Then, for each, calculate the economic strength $S = P^{1/3}G^{2/3}$, and $S_* = \log_e S$. Next, categorize the S_* values into a suitable class interval pattern and construct the histogram. Then compute the mean value $\overline{S}_*$ and the standard deviation σ_*.

Finally, now that you have determined the numerical values of $\overline{S}_*$, and σ_*, try to fit the normal probability distribution to your histogram. This distribution is given by equation (9.10).

For your information, the results of an analysis based on 1990 data are shown in figure 9.2. The normal distribution appears to fit the histogram fairly well. The economic strengths of the fifty states are listed in rank order in table 9.5.

10

How to Start Football Games, and Other Probably Good Ideas

We are now going to look at a few problems that are based on some simple concepts of probability. The first problem shows how you can determine the numerical value of $\pi = 3.1416$ by simply dropping a needle or a grain of sand onto a sheet of paper laid flat on a table. The second problem gives you the equation for designing a coin that, when tossed, is just as likely to land on its edge as it is on one side or another. The third problem produces a secret formula that will enable you to make millions, at meetings and parties, by wagering how many people in the group have the same birthday. The fourth problem shows how you can save a great deal of money in case you need to have all your teeth pulled out. Not only fascinating but also very useful information, right?

Buffon Needle Problem

This is a fairly ancient problem. It was formulated and solved by the French mathematician Count George de Buffon (1701–88). The problem is stated as follows. As shown in figure 10.1(a), a number of parallel lines are drawn on a plane surface a distance B apart. Then, at random, we drop a needle of length L, equal to or less than B, onto the surface. What is the probability, P, that the needle touches one of the lines?

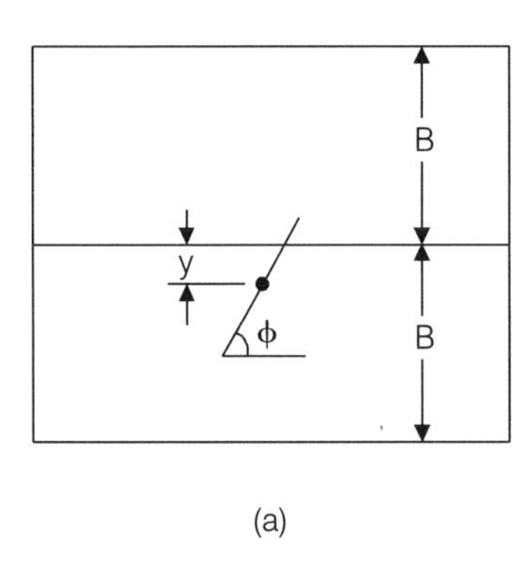

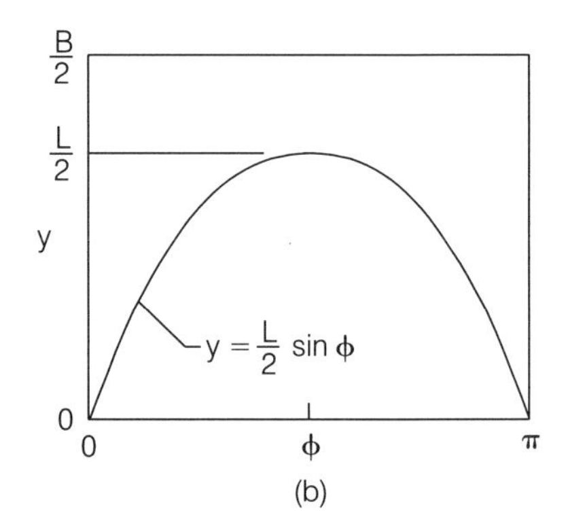

FIG. 10.1

The Buffon needle problem. (*a*) Definition sketch, with line spacing *B* and needle length *L*, and (*b*) rectangular and sinusoidal probability areas.

The distance between the center of the needle and the nearest line is y. The angle that the needle makes with the line is ϕ. The term "at random" means that any distance y is equally probable, that any angle ϕ is equally probable, and that y and ϕ are independent of each other.

From figure 10.1, we note that the needle will touch a line if

$$y < \frac{L}{2}\sin\phi. \tag{10.1}$$

Furthermore, y can have any value between 0 and $B/2$; ϕ can have any value between 0° and 180° or, better, 0 and π radians. That is,

$$0 < y < B/2; \qquad 0 < \phi < \pi. \tag{10.2}$$

Next, we look at the rectangle shown in figure 10.1(*b*). For any given toss, the independent random variables ϕ and y define a point, (ϕ, y), anywhere inside the rectangle. The area of the rectangle is $A = \pi B/2$.

However, equation (10.1) says that the needle will touch a line only if $y < (L/2)\sin\phi$. So in Figure 10.1(*b*) we draw the curve $y = (L/2)\sin\phi$. For a given toss, if the (ϕ, y) point falls *below* this curve, the needle has touched the line and we call the toss a "success." On the other hand, if (ϕ, y) falls *above* this curve, the needle has *not* touched the line and the toss is a "failure."

Well, what is the area of the "successful" region? To answer this question we need to solve the following very simple problem of integral calculus:

$$A_s = \int_0^{\pi} \frac{L}{2} \sin \phi \, d\phi = L. \tag{10.3}$$

So the "successful" area, A_s, is simply L. By definition, the probability of success, P, is the ratio of successful outcomes to all possible (success plus failure) outcomes. Therefore

$$P = \frac{A_s}{A} = \frac{L}{\pi B/2} = \frac{2L}{\pi B}, \tag{10.4}$$

and so

$$\pi = \frac{2L}{B} \frac{1}{P}. \tag{10.5}$$

We note that if $L = B$ then $\pi = 2/P$. It is hard to believe that such a difficult problem has such an easy solution.

All we need do now is to carry out some experiments; some rainy afternoon you might want to do this. Place a large sheet of paper on the dining room table and draw some parallel lines with spacing B. Next, start dropping a needle of length $L \leq B$ onto the sheet. Record the total number of tosses, T, and the number of successes, S. Then compute $P = S/T$ and, from equation (10.5), calculate the value of π.

Do not be discouraged if you do not quickly get the number 3.1416. This method of physically tossing a needle produces an answer that converges very slowly to the correct value. However, we can easily simulate this same problem on a computer and speed things up enormously. We will come back to this in a moment.

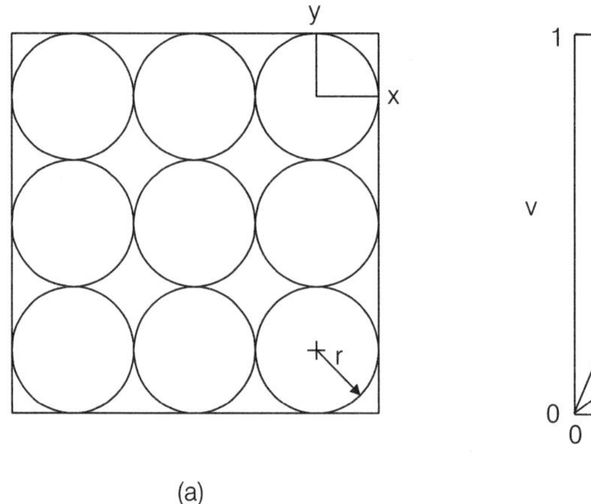

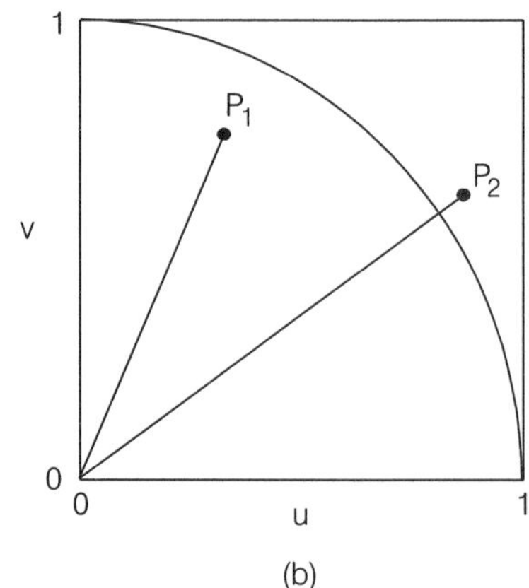

FIG. 10.2

The grain of sand problem. (*a*) An array of tangent circles of radius r. (*b*) Normalized probability areas with $u = x/r$ and $v = y/r$. For point P_1, $u^2 + v^2 < 1$, a "success," and for point P_2, $u^2 + v^2 > 1$, a "failure."

Grain of Sand Problem

A problem similar to the Buffon needle problem, and indeed even simpler, is one described by Rastrigin (1984) in his interesting little book. This problem involves dropping a grain of sand or a small bead onto a plane surface comprised of an array of tangent circles, as shown in figure 10.2(*a*).

Suppose that we have an array of 10 × 10 circles and that each circle has a radius r. Consequently, the area of the circumscribing square is $A = (10 \times 2r)^2 = 400r^2$. The total area of the 100 circles is $A_S = 100(\pi r^2)$.

Our experiments on the dining room table could not be easier. We simply drop a grain of sand from some height, perhaps equal to the length of the square, and record for each drop whether the sand grain landed inside a circle (a "success") or outside a circle (a "failure"). As in the needle problem, the success probability is

$$P = \frac{A_s}{A} = \frac{100\pi r^2}{400r^2} = \frac{\pi}{4}, \tag{10.6}$$

and so $\pi = 4P$. As before, from our experiments we get $P = S/T$ and from this we determine the value of π.

Again, a very simple answer. However, as before it will take a lot of sand grain dropping to get much beyond the 3 in π.

Forget it. Let us simulate the problem mathematically and then go to a computer. First, we select one-fourth of a circle of radius r, and bound it by a square. Then we "normalize" the rectangular coordinates, x and y, by dividing each by the radius, r. So we have $u = x/r$ and $v = y/r$.

Now we start our mathematical experiments using a table of random numbers, or even a hand calculator with a random number key, to get u and v and hence $u^2 + v^2$. With reference to figure 10.2(*b*), if $u^2 + v^2 < 1$, then mathematically our sand grain has landed at point P_1, which is *inside* the circle, and so the trial was a "success." If $u^2 + v^2 > 1$, the sand grain landed at point P_2, *outside* the circle, and so we have a "failure." Successive trials give us $P = S/T$ and hence $\pi = 4P$. What could be easier?

Well, the needle problem and the sand grain problem are simple examples of so-called Monte Carlo simulation. The methodology of this very powerful technique was developed in the late 1940s by mathematicians John von Neumann and Stanislaw Ulam, and has been greatly advanced since then. Monte Carlo methodology is extremely useful in applied mathematics, engineering, and science and has also been widely applied in economics, geography, psychology, and other branches of the social sciences. Suggested references are Rastrigin (1984) for an elementary presentation and Hammersley and Handscomb (1964) for a more advanced one. Most recent books dealing with the subjects of probability theory and operations research include topics on Monte Carlo simulation.

Thick Coin Problem

We have a coin of diameter D and thickness B, which we toss in the air. What is the probability that it will land on its edge? This is a generalization of a problem given by Mosteller (1965).

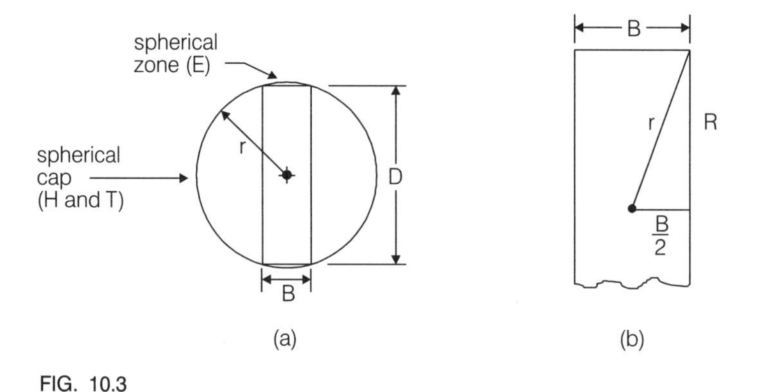

FIG. 10.3

The thick coin problem. (*a*) Sketch relating the coin to a concentric sphere and (*b*) sketch relating the coin and sphere dimensions.

As shown in figure 10.3(*a*), the easiest way to solve the problem is to imagine that the coin is embedded in a concentric sphere of radius r. The radius of the coin is $R = D/2$.

The probability that the coin will land on side H (heads) is proportional to the area of the *spherical cap* corresponding to side H. The same can be said about the probability of landing on side T (tails). The probability that the coin will land on the edge is proportional to the area of the *spherical zone* corresponding to edge E. Well, the area of the spherical zone is $A_s = 2\pi rB$ and the area of the entire sphere is $A = 4\pi r^2$. Accordingly, the probability of the coin landing on its edge is

$$P = \frac{A_s}{A} = \frac{2\pi rB}{4\pi r^2} = \frac{B}{2r}. \tag{10.7}$$

From figure 10.3(*b*), it is clear that $r = \sqrt{R^2 + (B/2)^2}$. Substituting this relationship in equation (10.8), setting $R = D/2$, and simplifying gives

$$P = \frac{1}{\sqrt{1 + (D/B)^2}}. \tag{10.8}$$

Solving for B/D gives the alternative result

$$\frac{B}{D} = \frac{P}{\sqrt{1 - P^2}}. \tag{10.9}$$

As we would expect, if $B/D \to 0$ (i.e., a very thin coin), then equation (10.9) gives $P = 0$. If $B/D \to \infty$ (a very thick coin), we get $P = 1$. In between these two extremes, we can easily calculate the "edge-landing probability" for any value of the thickness-diameter ratio, B/D. Table 10.1 gives a list of such probabilities for various objects.

By the way, in computing these values of P, we are interested only in the geometry we developed above. We are not interested in the dynamics of bodies in motion, so we ignore things like angular momentum, stability, moments of inertia, and the like.

Finally, if we want a coin that is equally likely to land on its edge, or on one side, or on the other side, that is, $P = 1/3$, then equation (10.9) says that the thickness-diameter ratio should be $B/D = 1/2\sqrt{2} = 0.354$.

What would be a practical application, you ask? Such a coin would certainly revolutionize football games. For example, at the opening ceremony, when the referee tosses our thick coin, if it

TABLE 10.1

Edge-landing probability P for various objects

Object	*B, cm*	*D, cm*	B/D	P
U.S. silver dollar	0.2	3.7	0.054	0.054
Roller skate wheel	1.0	5.0	0.20	0.196
Lifesaver candy	0.5	1.6	0.31	0.298
Can of tuna fish	4.0	8.0	0.50	0.447
Toilet paper roll (full)	12.0	12.0	1.00	0.707
Can of soup (any kind)	10.0	7.0	1.43	0.819
Toilet paper roll (empty)	12.0	4.0	3.00	0.949
Cigarette (unlighted)	8.0	0.8	10.00	0.995

comes up heads then Michigan kicks off. If it comes up tails, Notre Dame kicks. And if it lands on the edge, we start with marching and twirling!

Finally, dropping needles or grains of sand onto a sheet of paper to compute the value of π makes a great party game. So does tossing toilet paper rolls and cans of tuna fish to confirm edge-landing probabilities. Why don't you try it next time?

Birthday Problem

How many persons should there be in a group in order that there be a probability of 50% that at least two will have the same birthday (in a 365-day year)?

In problems like this, we first determine the probability Q that something will *not* occur. Then we use the relationship $P = 1 - Q$, for the probability that something *will* occur.

Let us proceed this way. There are n persons in a group. The probability is 365/365 that person number 1's birthday is on a certain date. The probability is 364/365 that person number 2's birthday is different from that of person number 1. The probability is 363/365 that person number 3's birthday is different from those of persons number 1 and 2, and so on. The combined probability that the n persons all have different birthdays is the product of the n individual probabilities. That is,

$$Q = \frac{365}{365} \times \frac{364}{365} \times \frac{363}{365} \times \cdots \times \frac{365 - (n - 1)}{365}. \quad (10.10)$$

Therefore, the probability that at least two persons do not have different birthdays, that is, do have the same birthday, is $P = 1 - Q$, where Q is given by equation (10.10).

From this equation and the relationship $P = 1 - Q$, we can easily compute the values of P for various values of n. The results are displayed in table 10.2. From the table, we see that if there are 22 or 23 persons in the group, there is about 50% probability that at least two will have the same birthday. If there are 40 persons, the probability is 89% and if there are 60, the probability is over 99%—a virtual certainty. Amazing, what?

TABLE 10.2

Probability P that at least two persons have the same birthday in a group of n persons

n	P	n	P	n	P
2	0.003	15	0.253	30	0.707
3	0.008	20	0.413	40	0.892
4	0.016	22	0.477	50	0.971
5	0.027	23	0.509	60	0.994
10	0.117	25	0.570	70	0.999

Incidentally, when the value of n in equation (10.10) is "large," the calculation of Q is tedious. In this case we can use the so-called Stirling approximation:

$$m! = \sqrt{2\pi m}\, m^m e^{-m}, \qquad (10.11)$$

where $m!$ is *m factorial.* For example, if $m = 5$, then by definition, $5! = 5 \times 4 \times 3 \times 2 \times 1 = 120$. With this approximation, equation (10.10) can be put into the form

$$Q = \left[\frac{365}{365 - n}\right]^{365-n+(1/2)} e^{-n}. \qquad (10.12)$$

This equation is surprisingly accurate; it is valid for values of $n \geq 5$.

Let us see how good our model is. Our test group is composed of the people who have been presidents of the United States. Through Bill Clinton, forty-one people have served (Grover Cleveland served two nonconsecutive terms). So, with $n = 41$, the probability is $P = 0.903$ that at least two presidents had the same birthday.

Sure enough! James K. Polk (eleventh president, 1845–49) was born on November 2, 1795, and Warren G. Harding (twenty-ninth president, 1921–29) was born on November 2, 1865. The fact that neither Polk nor Harding was among America's more spectacular presidents is not important. Indeed, we are in their debt because they make our mathematical model look good.

Teeth Extraction Problem

You remember Joe Smog. He was the guy in our skydiving class who made all kinds of silly mistakes. Well, not long ago, Joe began to have problems with his teeth. He went to see his dentist, who said, "Joe, it will be necessary to extract all of your teeth." So Joe replied, "Okay. How much will it cost?"

The dentist thought for a moment and then said, "I'm a sporting person, so I'll make a deal with you. I will extract all of your thirty-two teeth for a flat fee of $1 million or, if you prefer, charge you one cent for extracting the first tooth, two cents for the second tooth, four cents for the third tooth, eight cents for the fourth, and so on, up to and including the thirty-second tooth. You decide and let me know."

Joe went home and started computing. Now, Joe may not be the world's greatest skydiver but he's pretty swift with mathematics. He began by writing down the following numbers:

Tooth number:	1	2	3	4	5	6	7	8
Cost (cents):	1	2	4	8	16	32	64	128.

He quickly saw that the total cost could be given by the series

$$C = 2^0 + 2^1 + 2^2 + 2^3 + 2^4 + \cdots + 2^{31} = \sum_{j=1}^{32} 2^{j-1}. \tag{10.13}$$

He also quickly saw that he had a big task of calculating a lot of powers of 2 followed by a horrendous exercise in addition. So he devised the following formula to evaluate the arithmetic series:

$$C = \sum_{j=1}^{n} 2^{j-1} = 2^n - 1. \tag{10.14}$$

Then, with customary cleverness, he substituted $n = 32$ into this equation, divided by 100 (to convert cents to dollar), and got the answer $42,949,672.95.

Joe thought to himself, "No wonder the chap drives a Porsche and wears designer shoes. Look what he gets for extracting teeth from unsuspecting, gullible nonmathematicians." He further thought to himself, "What if I go to 2 dentists, each extracts 16 teeth, I pay $1,310.70; or 4 dentists, each extracts 8 teeth, I pay $10.20; or 8 dentists; each extracts 4 teeth, I pay $1.20; or 16 dentists, each extracts 2 teeth, I pay $0.48; or 32 dentists, each extracts 1 tooth, I pay $0.32?"

Well, Joe was about to select the least-cost strategy of $0.32, which meant going to 32 different dentists. However, at the last moment he said to himself: "Hold on! I must take the bus from one dentist to the next and each bus fare is $2.00. If I select the 32-dentist strategy, I will have to pay $64.00 in bus fares. I had better take this into account." So Joe did some more computing and produced table 10.3.

From the table, Joe concluded that, with bus fares taken into account, the least total cost would be $17.20. This would involve eight bus rides going to eight dentists with each dentist extracting four teeth.

TABLE 10.3

Results of computations for teeth extraction costs

Extracted by each dentist n	*Number of dentists* N	*Amount paid to each* $	*Cost for extraction* C_e, $	*Cost for bus rides* C_b, $	*Total cost* C, $
1	32	0.01	0.32	64.00	64.32
2	16	0.03	0.48	32.00	32.48
4	8	0.15	1.20	16.00	17.20
8	4	2.55	10.20	8.00	18.20
16	2	633.35	1,310.70	4.00	1,314.70
32	1	43×10^6	43×10^6	2.00	43×10^6

But Joe, being a differential calculus addict, was still not satisfied. He said to himself, "I recognize a mini-max problem when I see one. I shall examine this situation further." So he wrote down the equation

$$C = C_e + C_b = \frac{1}{100}(2^n - 1)\left(\frac{32}{n}\right) + 2.00\left(\frac{32}{n}\right), \qquad (10.15)$$

where C is the total cost, C_e is the teeth extraction cost, C_b is the bus ride cost, and n is the number of teeth extracted by each dentist. He then put equation (10.15) into the form

$$C = \frac{1}{n}(a2^n + b), \qquad (10.16)$$

where $a = 0.32$ and $b = 63.68$. To determine the premium cost, Joe first differentiated equation (10.16) and then set the result equal to zero to obtain

$$n_*^2 2^{n_* - 1} = 2^{n_*} + b/a, \qquad (10.17)$$

where n_* is the number of teeth extracted by each dentist to give the minimum total cost and $b/a = 199$. A few trial-and-error calculations with this equation gives $n_* = 4.47$. Substitution of this result into equation (10.16) yields $C_{\text{min}} = \$15.83$.

Well, Joe felt pretty good about getting this minimum-cost answer. Then he realized it would not be easy to take 7.16 bus rides to go to 7.16 dentists to have 4.47 teeth removed. So he said to himself, "I'd better go back to the $17.20 program." So he did.

11

Gigantic Numbers and Extreme Exponents

"The bigger the better, the more the merrier." These words seem to characterize the fascination we have about gigantic things and enormous numbers.

The Empire State Building is 1,250 feet high. The Queen Mary weighs 75,000 tons. The largest recorded iceberg covered an area of 10,000 square miles. The Panama Canal required the excavation of 250,000,000 cubic yards of rock. Over the past 1,000,000 years about 80,000,000,000 people have lived. The distance to the nearest star is 25,000,000,000,000 miles. And so on.

In this chapter we are going to look at a few situations involving awesomely large numbers. As a warming-up exercise, you can study the information presented in table 11.1, which lists several "items" and the "numbers" associated with them. You will agree that some of the items and numbers are pretty scary; others are simply depressing.

Rice Grains on a Chessboard: A Story in Two Parts

Part I

A long time ago there lived in India a wealthy and powerful maharajah. He possessed vast estates comprising endless acres of rice fields and countless numbers of gold mines. Nevertheless,

TABLE 11.1

Various items and their associated numbers in 1995, mostly in the United States

Item	*Number*
Total number of members of Congress	5.35×10^2
Number of deaths from car accidents per year	4.80×10^4
Number of violent crimes per year	1.65×10^6
Number of tons of wastes produced per year	2.65×10^8
Total number of people in the world	5.73×10^9
Number of hours Americans watch TV per year	6.25×10^{11}
Public debt of the federal government ($)	5.50×10^{12}
Energy consumption per year (BTUs)	8.50×10^{16}

even though he owned these fantastic assets and much more, his most treasured possession was his lovely young daughter.

One day, while jogging through her father's vast estates, the lovely daughter fell into a crocodile-infested river. Her cries of fear and anguish were heard by a handsome young man who, without hesitation, plunged into the river and brought the lovely maiden safely to shore only one foot ahead of a hungry crocodile.

The maharajah, upon learning of his lovely daughter's close escape from a horrible demise and the handsome young man's heroic feat, summoned the latter to his corporate headquarters.

"Young man," exclaimed the maharajah, "thou hast saved my lovely young daughter from a horrible demise. I shall reward thee. What dost thou seek to possess?"

"Oh sire!" replied the handsome young man, "I would cherish a chessboard. You know, standard 8 × 8 matrix. Gold preferably, otherwise three-ply."

"That will be no problem," responded the maharajah, "I am the owner of most of the gold in India. You'll have your board first thing in the morning. What else?"

"Oh sire!" said the handsome young man, "I love rice. I find it tasty, it is low in fat and cholesterol, and it satisfies my minimum daily requirements for vitamins and stuff. Therefore, I humbly request that you instruct your rice foreman to place one grain of rice on the first square of my chessboard, two grains on the second square, four grains on the third, eight grains on the fourth, sixteen on the fifth, and so on, to and including the sixty-fourth square. I seek nothing further, sire."

"Big deal!" chortled the maharajah. "Your request, though pathetically trivial, is granted. I must depart now. Thanks again for saving my lovely young daughter from the crocodile-infested river."

Well, it turns out, and unbeknownst to the maharajah, the handsome young man was actually a math major at the nearby college. So he knew, as we learned in chapter 10, that the total number of grains of rice on the chessboard could be computed from the equation

$$N = 2^n - 1. \tag{11.1}$$

Consequently, with $n = 64$, he determined that the total number of rice grains on the chessboard would be

$$N = 18{,}446{,}744{,}073{,}709{,}551{,}615,$$

which is a considerable amount of rice. This large number can be written as $N = 1.845 \times 10^{19}$. Let's play around with it a bit.

Volume of the Rice Grains

There are about 15×10^6 grains of rice in a cubic meter. So the total volume of the rice on the chessboard is $V = (1.845 \times 10^{19})/(15 \times 10^6) = 1.23 \times 10^{12}$ m^3. If all this rice were placed in a cubical box, the sides of the box would have a length of $L = V^{1/3} = 10{,}700$ m $= 10.7$ km.

The area of the continental United States is $A = 7{,}616{,}000$ km^2. If the rice were spread evenly over the entire U.S., it would be $H = V/A = 0.16$ m $= 16$ cm $= 6.3$ in deep.

Total Length of the Rice Grains

The length of a grain of rice is about 4 mm $= 4 \times 10^{-6}$ km. If all this rice were placed end to end, the total length would be

$L = 7.38 \times 10^{13}$ km. Now light travels at a velocity of about 300,000 km/s or 9.46×10^{12} km/yr. So the length of the rice train would be approximately 7.8 light years; this is nearly double the distance to the nearest star, Proxima Centauri.

Weight of the Rice Grains

There are about 45,000 grains of rice in one kilogram. So the total weight of our rice grains is $W = 4.1 \times 10^{14}$ kg $= 4.1 \times 10^{11}$ metric tons = 410,000 million tons. The total worldwide annual production of rice is around 300 million tons. Based on this present rate, our chessboard rice represents approximately 1,365 years of rice production. It is safe to say that this is more rice than has been grown worldwide since the dawn of civilization.

Nutritional Value of the Rice Grains

Demographers have determined that since the dawn of mankind, say, over the past million years, approximately 80×10^9 people have been born and lived an average of twenty-five years. On this basis, each one gets about 5,100 kilograms from our enormous pile of rice; this comes to around 0.56 kilograms per day for everybody.

It turns out that one kilogram of rice contains 3,500 calories. So everyone receives about 2,000 calories per day, which is enough for survival. The drawback, of course, is that most of these eighty billion people would probably get sick and tired of eating nothing but rice.

Rice Grains on a Chessboard: A Story in Two Parts

Part II

The grateful maharajah ordered his rice foreman to begin delivery of the rice to the handsome young man. The first row ($n = 8$) comprised 255 grains and the second row ($n = 16$) brought the total to 65,535, or about 1.45 kilograms. The third row (n = 24) yielded a total of 373 kilograms, so they started putting the rice into 100-kilogram sacks. The fourth row (n = 32) required 954 sacks, each one weighing 100 kilograms. Since they were being delivered by the foreman at the rate of one sack per second, this required 15.9 minutes.

By this time, the maharajah could see how things were going. He had stopped laughing and was getting quite pale. Halfway through the fifth row (n = 36) with 15,270 sacks and 4.24 hours delivery time, he knew that he had been aced.

At the end of the fifth row (n = 40), with a total of 244,336 sacks and a delivery time of nearly sixty-eight hours, the maharajah was desperate; he ordered his lovely young daughter thrown back into the crocodile-infested river. Midway through the sixth row (n = 44), with a total of 3.91 million 100-kilogram sacks and a delivery time of forty-five days, he himself jumped into the river. (You can do the rest of the squares. Just delivering the last square's rice took sixty-five thousand years.)

The handsome young man, though grief-stricken about the loss of the lovely maiden and her generous father, chuckled, "Had he taken 'Math 101: Probability and Statistics,' as I did, he would have been sufficiently swift to have ordered me to go get my own rice out of the warehouse. I would have given up at about n = 28. Ah, cruel fate."

The Temple of Hanoi in Benares

Before we leave India, here is another classic big-number problem. It is reputed that long ago, at the Temple of Hanoi in the holy city of Benares on the Ganges river, the following question was posed.

There are three small-diameter vertical rods. The left one contains sixty-four rings with the largest-diameter ring on the bottom and ever-decreasing diameter rings stacked upward. The problem is to transfer all sixty-four rings to the rod on the right and end up with the same stacking pattern as that at the outset. The middle rod is used as a stopping place. At no time is a ring allowed on top of a ring of smaller diameter. How many moves are required?

Interestingly, the answer is the same as for the chessboard problem: $2^{64} - 1$. At the rate of one move per second, it would take 585 billion years to complete the transfer. It is noted that the earth is only about five billion years old and it will probably be around for only another five billion. So it's too late to start the exercise. Of course, we could speed up the problem with a superfast computer, but it would still take a long time.

Drops of Water in the Ocean

Now here is a problem quite similar to the rice grain problem. The volume of all the water in the world—oceans, ice caps and glaciers, groundwater, lakes and rivers—is approximately $V = 1{,}387.5 \times 10^{15}$ m^3. Assume that a drop of water has a diameter of 3 mm, so its volume is $v = (\pi/6)(3)^3 = 14.14 \times 10^{-9}$ m^3. Consequently, the number of drops comprising all the world's water is $N = 9.815 \times 10^{25}$.

Okay. Here is a problem you can do. We are going to put all of these drops into containers of various sizes. However, instead of *doubling* each time, as we did in the rice problem, we are now going to *triple*. That is, there will be one drop in container 1, three drops in container 2, nine drops in container 3, 27 drops in container 4, and so on.

This time, the equation for the total number of drops is

$$N = \tfrac{1}{2}(3^n - 1). \qquad (11.2)$$

PROBLEM 1. Show that fifty-five containers will hold most of the water drops.

PROBLEM 2. The containers are cubically shaped. Confirm that the side length of the fifty-fifth cube is about 940 kilometers.

PROBLEM 3. The total volume of the five Great Lakes is 22,650 km^3. Show that the first forty-five containers are needed just to hold the water drops comprising the Great Lakes.

That's enough. We stop here.

Chimpanzees on Typewriters

It was a dark, stormy night as the late-shift research assistant made his way along the east wing of Research Building 192. He stopped at Work Station No. OE97-41-921A and observed the printout coming from the computer console. For a moment he was absolutely aghast as he read:

To be or not to be, that . . .

He hit the alarm bell; the research supervisor arrived instantly. Together they read the rest of the-randomly typed text of the chimpanzee:

To be or not to be, that is the phlugdop mizburp

"Wow!" exclaimed the supervisor, "That was indeed a close one." They both went back to work.

At this point we pause to acknowledge Edward George Bulwer-Lytton (1803–73), English novelist, for his incredibly unique way to start a story ("It was a dark and stormy night . . . "), and William Shakespeare (1564–1616), English poet and playwright, for providing the several words ("To be or not to be . . . ") from Act III of *Hamlet*.

We also acknowledge good ole Snoopy in *Peanuts* for his unfailing loyalty to Bulwer-Lytton concerning the proper way to begin a melodrama, and comedian Bob Newhart for setting the stage, here elaborated, for the experimental proof of the long-held theory that, given a sufficiently large number of chimpanzees and an equally large number of typewriters, the complete works of Shakespeare could be produced by random typing.

Typing the Alphabet

Well, what *is* the probability that the Bard's complete works could be generated by a random process, such as chimpanzees on typewriters; or even *Hamlet* or even Act III of *Hamlet* or even

To be or not to be, that is the question,

or even

To be or not to be...

Counting spaces and the comma, there are forty characters in the first line and eighteen in the second. As we shall see, it is quite unlikely that even these simple word clusters could be generated, in a reasonable period of time, by battalions of speedy randomly typing chimpanzees.

Let us start with something easier. Suppose we simply want to reproduce the alphabet in proper order by hitting the typewriter —sorry, console—keys at random. In this case, we do not worry about the space key nor any punctuation marks; we are concerned only with striking 26 letter keys 26 times.

The probability of typing the *a* as the first letter is $(1/26)$. The probability of typing the sequence *ab* is $(1/26)(1/26)$. The probability of typing the sequence *abc* is $(1/26)(1/26)(1/26) = (1/26)^3$. Consequently, the probability of typing the entire alphabet *abc... xyz* is $(1/26)^{26}$.

We let $p = 1/26^{26}$ and $q = 1 - p$, where p is the probability of success of a trial and q is the probability of *failure*. By a *trial* we mean one twenty-six-letter typed sequence. It is surely obvious that $p = 1/26^{26}$ is an extremely small number or, said another way, 26^{26} is an extremely large number. In fact, $26^{26} = 6.156 \times 10^{36}$ is a number nearly a billion billion times larger than the number of rice grains on our chessboard.

Okay. We now commence our experiments with a series—a very long series, it turns out—of repeated trials. Such repeated independent trials, with constant values of p and q, are called

Bernoulli trials. From virtually any book dealing with probability theory, for example, Rozanov (1977), we obtain the equation

$$P(k) = C(n,k)p^k q^{n-k}, \tag{11.3}$$

where

$$C(n,k) = \frac{n!}{(n-k)!k!}. \tag{11.4}$$

The quantity $P(k)$ is the probability that, with n independent trials, a success occurs exactly k times; p is the probability of success in a single trial and q is the probability of failure in a single trial. The expression given by equation (11.3) is called the *binomial distribution.* The relationship of equation (11.4) gives the number of combinations of n things taken k at a time. The symbols with the exclamation marks, you will recall, are called factorials. For example, $5! = 120$ and $10! = 3{,}628{,}800$.

Let us look at a simple problem. In the roll of a die, what is the probability of getting at least one 6 in n rolls? Well, in a single roll or trial, the probability of success (i.e., getting a 6) is $p = 1/6$, and so the probability of failure (not getting a 6) is $q = 1 - p = 5/6$. From equation (11.3), the probability of getting exactly k successes in n trials is then

$$P(k) = C(n,k)\left(\tfrac{1}{6}\right)^k \left(\tfrac{5}{6}\right)^{n-k}. \tag{11.5}$$

Therefore, the probability of getting no successes ($k = 0$) in n trials, with $C(n,0) = 1$, is $P(0) = P_0 = (5/6)^n$. Consequently, the probability of getting at least one success in n trials is $Q_0 = 1 - P_0 = 1 - (5/6)^n$. If $n = 1$ then $Q_0 = 0.167$; if $n = 5$, $Q_0 = 0.598$; if $n = 10$, $Q_0 = 0.838$; and so on. We would need about sixteen trials to be 95% certain of getting a six.

We solve our chimpanzee problem the same way... well, almost. It is not practical to use the binomial distribution in this case, because p is extremely small and n is extremely large. The computations would be horrendous. Fortunately, however, if the product np is a small number, say about 5 or less, then the binomial distribution reduces to the *Poisson distribution*

$$P(k) = \frac{r^k}{k!}e^{-r}, \tag{11.6}$$

where $r = np$.

The rest is easy. The probability of exactly zero successes in n trials is $P(0) = P_0 = e^{-r}$. Therefore the probability of at least one success is $Q_0 = 1 - P_0 = 1 - e^{-r}$.

Now suppose we want to be 99% certain of getting at least one successful alphabet typing in n trials. Then $Q_0 = 0.99$, $P_0 = 0.01$, and so $r = np = 4.605$. Since we already know that $p = 1/(6.156 \times 10^{36})$, then $n = 2.835 \times 10^{37}$. This result indicates that with this many trials, we can be 99% certain that the chimpanzees will correctly type the alphabet. Of course, with pretty good luck, one of them could do it the first time!

Want Some More Big Numbers?

We now put 35 keys on the console to provide a space key and punctuation marks; we forget about capitalizing letters.

Our friend Snoopy would need about two billion typings just to get his name right.

"To be or not to be"
requires around 6×10^{27} trials.

"To be or not to be, that is the question"
will need 6×10^{61} attempts.

There are approximately 250,000 characters in *Hamlet*; in your spare time, you may want to confirm this. Therefore, about $10^{386,000}$ (which is 10 followed by 386,000 zeros) trials will be necessary to come close to a random typing of the entire play. This number is so big, we can even go to a second story:

$$10^{386,000} = 10^{10^{5.587}}.$$

Using no more than three digits, what is the largest number one can devise in decimal notation with no additional symbols? The answer is

$$9^{9^9} = 10^{3.697 \times 10^8} = 10^{10^{8.5678}}.$$

Five Lively References, Two Without Mathematics

Anyone interested in large numbers—and that includes most of us—should read the very interesting book by Davis (1961) entitled *The Lore of Large Numbers*. In the same category of pleasant reading is the book by Perelman (1979) called *Mathematics Can Be Fun*. He presents the "rice on a chessboard" problem and a great many other interesting, indeed amusing, situations involving mathematics.

In his intriguing book on problem-solving with computers, Bennett (1976) devotes about one hundred pages to some of the mathematics associated with language and linguistics. He presents applications in the fields of anthropology, communications, cryptography, and information theory. He also deals with "monkeys at typewriters," but delves much deeper into the problem than we have.

An excellent reference for everyone interested in mathematics is the four-volume set *The World of Mathematics* edited by Newman (1956). Surprisingly, it contains very little mathematics; it's easy to read. In the section "Mathematics in Literature," there is a clever story entitled "Inflexible Logic" by Russell Maloney, who was a writer for the *New Yorker* for a long time.

In the story, Mr. Bainbridge acquired six chimpanzees in order to test the theory about random typing. It turns out that the six chimps were fantastically lucky, so lucky, in fact, that they succeeded in reproducing two or three dozen literary classics without any mistakes. Tragically, these entirely unreal accomplishments by the chimpanzees drove Professor Mallard, the mathematics consultant, completely crazy. So he plugged Mr. Bainbridge and the chimps. As he fell to the floor, however, Bainbridge got off a lucky shot. That was the end of Mallard and the tale.

A short story entitled "Epicac" appears in the collection *Welcome to the Monkey House* by Kurt Vonnegut (1970). Epicac is the name of a supercomputer owned and operated by the military. Our hero, one of the programmers using Epicac to solve important military problems, is having zero success in persuading the fair maiden, also a programmer, to become his wife. Our hero, however, discovers she is a pushover for romantic poetry. So he programs Epicac to write beautiful stuff—in much the same way and with the same degree of success as Bainbridge's incredibly lucky chimpanzees. As fate would have it though, Epicac also falls in love with the maiden. However, knowing that it was never meant to be, Epicac commits suicide and leaves only smoking wires and melted transistors to reveal where he once was. Needless to say, the general is very upset. At the end, our hero and the fair maiden disappear into the sunset looking for a cottage and new jobs.

12

Ups and Downs of Professional Football

"The rich get richer and the poor get poorer." That may be true for a lot of things but it's certainly not true as far as team performance in the National Football League is concerned. We shall come back to this point later on.

In chapter 2, we introduced the following simple relationship, which describes so-called Malthusian or exponential growth:

$$\frac{dN(t)}{dt} = aN(t), \tag{12.1}$$

where $N(t)$ is the magnitude of something at time t, and a is a growth coefficient. This equation says that the *rate* at which the magnitude of the "something" is increasing is directly proportional to its present magnitude. The "something" could be the number of alligator eggs in Florida, the number of dollars the federal government has managed to borrow, the number of cockroaches in your kitchen, or the amount of money you have in your savings account.

The solution to equation (12.1) is

$$N = N_0 e^{at}, \tag{12.2}$$

in which N_0 is the magnitude of N at time $t = 0$.

This is the well-known exponential growth equation. It says that N increases with time without limit. As t goes to infinity, N goes to infinity. We easily determine that the time it takes to double the magnitude of N is $t_2 = \log_e 2/a \doteq 70/a(\%)$. For example, if $a = 7.0\%$ per year, then N will double in about ten years.

At the other extreme we have the relationship

$$\frac{dN(t)}{dt} = -aN(t). \tag{12.3}$$

This expression indicates that the rate at which something is decreasing is directly proportional to its present magnitude. The solution is

$$N = N_0 e^{-at}. \tag{12.4}$$

This is the equation for exponential decay. In this case, the "something" could be grams of a radioactive substance, the concentration of a biological waste in a river, the intensity of a light beam in a thick fog, or the temperature of your cup of coffee. Again, the half-length time is $t_{1/2} \doteq 70/a(\%)$. We note that, with decay, as t goes to infinity, N goes to zero.

Wherein We Talk about the Rich Getting Richer and the Poor Poorer

Here is a model we look at initially. Suppose there are two colleges, one with a fairly good football team (college A) and the other with a fairly bad team (college B).

To simplify our model and illustrate our point, we disregard all factors except the following. The best of this year's crop of football players graduating from high school go to college A where there is already a "good or very good team." On the other hand, the worst of this year's graduating crop go to college B where there is already a "bad or very bad team." To make the obvious point: with passing time, the good team at college A gets better and then excellent and the bad team at college B gets worse and then horrible. This idealized model illustrates the tendency for the "rich to get richer" (exponential growth; college A) and the "poor to get poorer" (exponential decay; college B).

This scenario, and the associated human psychology involved, can be extended to basketball, swimming, and tennis teams. And to university departments of history, physics, and sociology. And to which specialized branch of the armed forces you want to serve in and which company you wish to work for. And to which neighborhood you seek to live in and which college you plan to send your children to. And to life in general.

We are getting off the subject. All we want to do is to take a look at professional football and carry out some mathematical analysis. Let's move on.

The Stabilizing System of Professional Football

This business of going either to infinity with the exponential growth of equations (12.1) and (12.2) or to zero with the exponential decay of equations (12.3) and (12.4) would, of course, be catastrophic in professional football. How much duller could it get: the Raiders win all their games forever and ever and the Buccaneers lose all of theirs from here to eternity. Fortunately, many years ago, the National Football League figured out how to avoid the problem; we shall go into detail about that in a moment.

First, we present the following relationship to replace the ones we had before, equations (12.1) and (12.3):

$$\frac{dU(t)}{dt} = a[U_m - U(t)]. \qquad (12.5)$$

This is a first-order linear differential equation. Here, U is the decimal fraction of games won by a NFL team during a particular season. For example, in 1986 the San Francisco Forty-Niners won ten games, lost five, and tied one. So $U = (10 + 0.5)/16 = 0.656$. Obviously, U must be between zero and one, that is, $0 \leq U \leq 1$. In equation (12.5), t is time in years; a is a growth coefficient,

1/years; and U_m is the League-wide average value of U: $U_m = 0.500$.

Now equation (12.5) says that the rate at which U changes at any time t is proportional to the difference between U_m and the value of U at that same time t. For a particular team, suppose that $U < U_m$. Then $dU/dt > 0$ (i.e., positive slope) and the team's performance is improving. On the other hand, suppose that $U > U_m$. Then $dU/dt < 0$ (negative slope) and the team's performance is deteriorating. We are tempted to conclude that this mathematical model helps: the "poor get richer and the rich get poorer."

But look what happens now. The solution to equation (12.5) is

$$U = U_m - (U_m - U_0)e^{-at}, \tag{12.6}$$

where U_0 is the value of U at $t = 0$ (say, at the end of the 1960 season). However, notice that as $t \to \infty$ then $U \to U_m$ regardless of the value of U_0. This means that ultimately the records of all teams in the League monotonically (i.e., smoothly and continuously) approach the value $U_m = 0.500$. There will be no ups and downs. Sooner or later, all teams would end up with win:loss:tie seasons of 8:8:0 or even worse, 0:0:16. How exciting could it get? Why stay up late on Monday nights? How would they decide who goes to the Super Bowl?

How do we get out of this foolish situation? Well, expressed in mathematical terms, here is what the National Football League did a long time ago:

$$\frac{dU(t)}{dt} = a\left[U_m - U(t - \tau)\right], \tag{12.7}$$

which is identical to equation (12.5) except that we have introduced a delay time, τ, in the argument on the right-hand side. This is a first-order linear *discrete-delay* differential equation.

In this form, the equation says that the rate at which U changes at the present time t is proportional to the difference between U_m and the value of U at some previous time $t - \tau$. For example, suppose that τ years ago (e.g., 2 or 3), team A had an excellent season record of $U = 0.90$. Then, taking $U_m = 0.50$ (and, say, $a = 1$), the present value of dU/dt would be -0.40; the negative sign means its present performance is getting worse. On the other hand, suppose that τ years ago team B had a dismal $U = 0.10$. Then the present magnitude of dU/dt would be $+0.40$; the positive sign indicates that its present performance is improving.

The outcome of this mechanism is that good teams become worse and poor teams become better, but not through the deadening route given by equations (12.5) and (12.6). As we shall see, a nice feature of the discrete-delay equation (12.7) is that its solution provides *oscillations* of the values of U.

This desirable "up and down" pattern of team performance is accomplished by the draft choice procedure of the National Football League. Overly simplified, the order of selection of new talent by NFL teams is in reverse order to the performance ranking of the previous season. Not surprisingly, it takes a certain amount of time for a team to turn around—for better or worse. This is one of the things we are after: what is the magnitude of the turnaround or delay time, τ?

If everything were that simple, data analysis of NFL scores would surely yield quite precise values of delay times. However, it is obvious that a lot of other factors are in the picture. For example, what are the effects of appointment of a new manager or coach, injuries to a few key players, shrewd or not-so-shrewd player trades, financial crises, move to a new stadium or a new city, unexpected good luck or bad luck, flukes, gremlins, and so on?

In other words, the "feedback signal"—the worst teams drafting the best players—is accompanied by numerous other signals: all the above factors, including the flukes and gremlins. Very complicated. However, in spite of all this, we will do the best we can to determine the up-down-up times, that is, the oscillation periods, T, and the turnaround times, that is, the delay times, τ.

How to Get a Mathematical Solution to a Delay Equation

Would you like a fairly easy problem in mathematics? You may have equation (12.5). This is the kind of problem you get quite early in your course in elementary differential calculus. The answer, of course, is equation (12.6).

Would you like a quite difficult problem in mathematics? Please take equation (12.7). This is the kind of problem you don't even see until you are a graduate student in mathematics, physics, or engineering; you may not see it even then. Isn't this unexpected? The inclusion of a simple little delay time, τ, turns a simple problem in mathematics into a very fierce one. Strange!

There are several ways to solve equation (12.7). The most straightforward method utilizes the so-called Laplace transformation. Another way is the method of steps. If you want to look into the matter, a good place to start is the *Handbook of Differential Equations* by Zwillinger (1989). He gives a number of references, including Bellman and Cooke (1963) and Saaty (1981). If you are interested in learning more about delay equations, you might want to consult MacDonald (1989); he gives quite a few examples and a lot of suggestions about other sources of information.

There is a fairly easy path to take to obtain a solution to equation (12.7). It can be a little risky because it introduces an infinite series that does not always converge. However, we shall assume that no such difficulty arises in our problem.

Here is our method. First we use Taylor's series,

$$f(x+s) = f(x) = \frac{1}{1!}s\frac{df}{dx} + \frac{1}{2!}s^2\frac{d^2f}{dx^2} + \cdots. \qquad (12.8)$$

This equation gives the value of a quantity f at a point, $x + s$, when we know its value at a nearby point x. If we apply this relationship to the $U(t - \tau)$ term on the right-hand side of equation (12.7), we get

$$U(t-\tau) = U(t) - \tau\frac{dU(t)}{dt} + \frac{1}{2}\tau^2\frac{d^2U(t)}{dt^2} - \cdots. \qquad (12.9)$$

Terminating the series at the end of the third term, substituting the result into equation (12.7), and carrying out some algebra, we obtain

$$\frac{d^2U}{dt^2} + \frac{2(1 - a\tau)}{a\tau^2}\frac{dU}{dt} + \frac{2}{\tau^2}U = \frac{2}{\tau^2}U_m. \tag{12.10}$$

This is what they call a second-order linear differential equation. It's a bit harder to solve than equation (12.5), but a good student in calculus could promptly give you the following answer:

$$U = U_m + e^{-pt}(C_1 \cos \omega t + C_2 \sin \omega t), \tag{12.11}$$

in which

$$p = \frac{1 - a\tau}{a\tau^2}; \qquad \omega = \sqrt{\frac{2}{\tau^2} - \left(\frac{1 - a\tau}{a\tau^2}\right)^2} \tag{12.12}$$

and C_1 and C_2 are constants. In the trigonometric terms of equation (12.11), ω is the *frequency* of oscillation. The *period* of oscillation is given by the relationship $T = 2\pi/\omega$.

Looking ahead, the two things we want to determine are the period, T, between high values or between low values of U of a particular NFL team, and the delay time, τ, for a team to turn around.

Now look at the quantity $\exp(-pt)$ in equation (12.11) and the definition of p in equation (12.12). In the latter, if $a\tau < 1$ then p is positive, and so, according to equation (12.11), U will "attenuate," that is, it eventually becomes zero but oscillates as it does so. On the other hand, if $a\tau > 1$, then p is negative, and so U will again oscillate but also "amplify," that is, it goes off to infinity. Incidentally, in our problem, U can never be greater than one nor less than zero.

Well, what we want is an oscillation that neither attenuates nor amplifies. So we set $p = 0$ and so $a\tau = 1$, $\omega = \sqrt{2}/\tau$, and $\tau = T/\pi\sqrt{2}$. Consequently, equation (12.11) becomes

$$U = U_m + C_1 \cos \frac{2\pi}{T}t + C_2 \sin \frac{2\pi}{T}t, \tag{12.13}$$

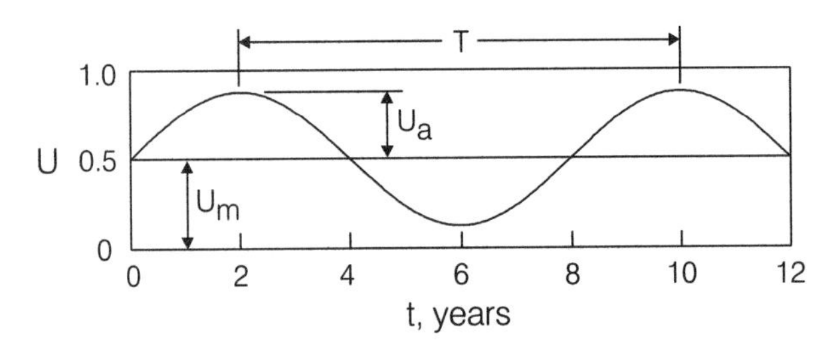

FIG. 12.1

Plot of an idealized "up and down" pattern of a NFL team's performance. For this curve, $U_m = 0.50$, $U_a = 0.40$, and $T = 8$.

where, again, C_1 and C_2 are constants. Without loss of generality, as the mathematician likes to say, equation (12.13) can be written in the form

$$U = U_m + U_a \sin \frac{2\pi}{T} t, \tag{12.14}$$

where U_a is the amplitude of oscillation and U_m (the League-wide average) is the value of U at $t = 0$. A plot of equation (12.14) is given in figure 12.1, with $U_m = 0.50$, $U_a = 0.40$, and $T = 8$ yr.

Oscillation Periods and Delay Times of NFL Teams

Football experts seeking to learn some mathematics and mathematics experts seeking to learn some football should study the following sections. The methodology of our analysis is given.

First, using the excellent information source *The Sports Encyclopedia: Pro Football* by Neft et al. (1992), the values of U were calculated for all teams of the National Football League for the period from 1960 through 1992. Second, this information, involving thirty-three seasons and twenty-eight teams—fourteen in the American Football Conference (AFC) and fourteen in the National Football Conference (NFC)—was plotted with time t (years) and annual performance U as coordinates. Third, from these plots, the periods of oscillation, T, were measured for each team; all data were analyzed to obtain the average value of T for each. This information was then used to determine the mean value, $\overline{T}$, and the standard deviation, σ, for the AFC and the NFC. These results were combined to yield the values of $\overline{T}$ and σ for the entire League. Finally, knowing the value of the period of oscillation, T, it was possible to compute the delay time, τ, from the equation $\tau = T/\pi\sqrt{2}$; the growth coefficient, a, was calculated from the expression $a = 1/\tau$.

Here are the main results. Two typical plots of the basic U versus t data are shown in figures 12.2 and 12.3, the former for the AFC's Buffalo Bills and the latter for the NFC's Chicago Bears. A word of explanation. The small open circles connected by dashed lines show the actual *annual* values of U. However, to simplify things, a decision was made to base the analysis on *biennial* values. This procedure tended to smooth out the signal without losing much information. The small solid circles connected by solid lines show the biennial data.

Period of Oscillation, T

Along the top of figures 12.2 and 12.3 are numbers showing the time, in years, between the peaks of the $U(t)$ plots. For example, for the Bills, we read 10, 6, and 10. Similar numbers along the bottom of the figures give the times between the *troughs*. For the

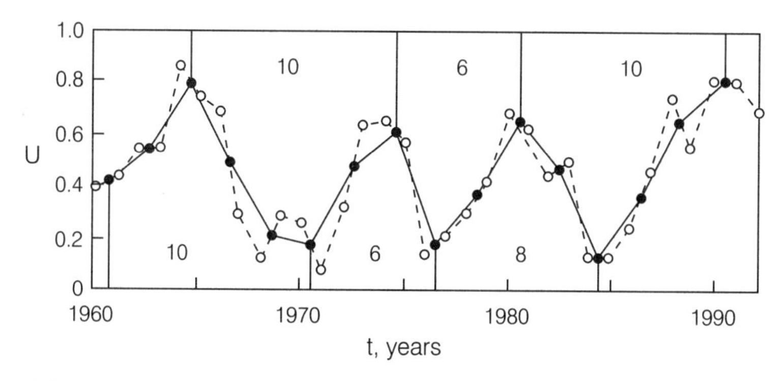

FIG. 12.2

Performance record of the Buffalo Bills, 1960 through 1992. Numbers along the top and bottom of the plot indicate the time, T, between peaks and between troughs.

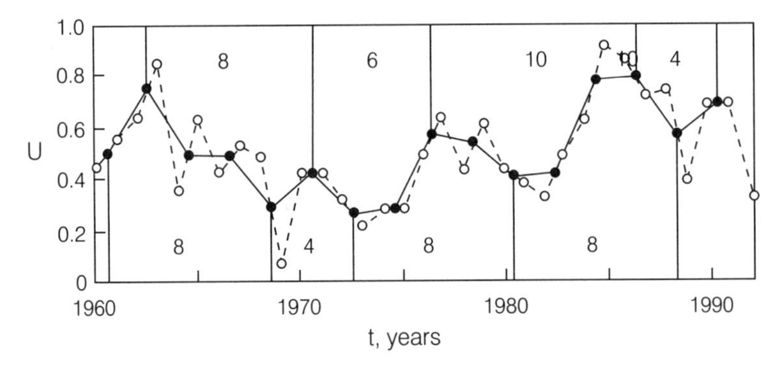

FIG. 12.3

Performance record of the Chicago Bears, 1960 through 1992. Time intervals $T \leq 4$ were discarded from the analysis.

Bills, we have 10, 6, and 8. The average value of these six numbers gives, for the Buffalo Bills, $T = 8.3$ yr. For the Chicago Bears, we get $T = 8.0$ yr.

Similar analyses were made of the records of all the other NFL teams. In several instances, times between peaks or between troughs were four years or less. These short time interval data seemed to provide little or no information; they were discarded.

Next, the measured values of T, obtained from the $U(t)$ plots of all fourteen teams of the AFC were combined, and the mean value, $\overline{T}$, and standard deviation, σ, were computed. Similar calculations were made to obtain the values of $\overline{T}$ and σ for the NFC. Finally, the values of T for all twenty-eight teams were utilized to provide $\overline{T}$ and σ for the entire NFL. All of these results are shown in table 12.1.

The histogram of the T data of the entire league is displayed in figure 12.4. The mean value $\overline{T} = 8.24$, the standard deviation $\sigma = 2.19$, and the coefficient of variation C.V. $= \sigma/\overline{T} = 0.266$. A linear distribution seems to fit the data reasonably well. The equation of the distribution is

$$N = \frac{2N_0T_b}{(T_b - T_a)^2}\left(1 - \frac{T}{T_b}\right), \tag{12.15}$$

TABLE 12.1

Computation results to determine periods of oscillation of performance cycles

Conference / League	*Mean value* $\bar{T}$, *years*	*Standard deviation* σ, *years*
American Football Conference	8.61	2.23
National Football Conference	7.88	2.14
National Football League	8.24	2.19

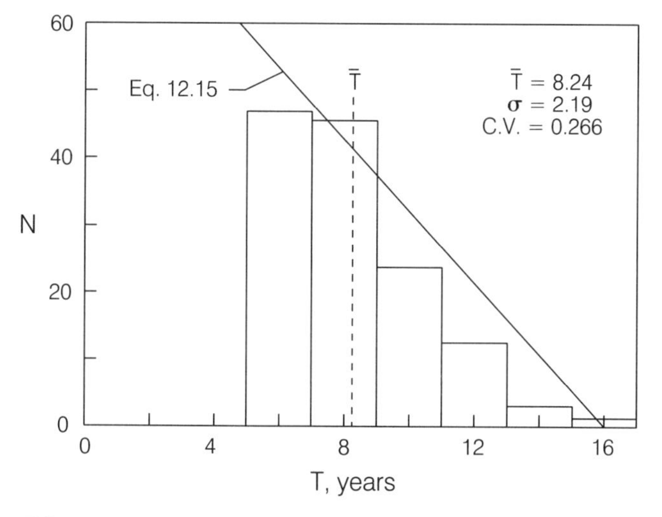

FIG. 12.4

Histogram of the oscillation periods T of the twenty-eight National Football League teams. $N_0 = 134$, $T_a = 6$, and $T_b = 16$.

in which, for our analysis, $T_a = 6$ and $T_b = 16$; the total number of measured values of T was $N_0 = 134$.

Delay Time, τ

Earlier we obtained the equation relating the delay time, τ, to the period of oscillation, T, that is, $\tau = T/\pi\sqrt{2}$. Substituting the League average, $T = 8.24$, into this equation gives $\tau = 1.85$ yr.

On this point, it is important to mention the following. Recall that we obtained an approximate solution to equation (12.7) by using Taylor's series; this yielded the result we just now employed, $\tau = T/\pi\sqrt{2}$. However, an *exact* solution to equation (12.7), corresponding to $p = 0$, provides the answer $\tau = T/4$. With $T = 8.24$ yr, this relationship gives $\tau = 2.06$ yr.

So here are our main conclusions. The average period of oscillation (i.e., from peak-to-peak or trough-to-trough) is approximately $T = 8$ yr. In turn, the average delay time (i.e., "turn around" time) is about $\tau = 2$ yr.

Growth Coefficient, a

For our purpose, we do not need to know the value of the growth coefficient, a. However, since it is easily available, we will calculate it. Recall that we set $p = 0$ in equation (12.11); this gave the result $a\tau = 1$. With $\tau = 2$ yr, we obtain $a = 0.5$ per year.

An Example of the Method of Computation

To illustrate the methodology we have developed, here is a simple example. It involves the Buffalo Bills.

In 1984, the Buffalo Bills completed the season with a grim (2:14:0) $U = 0.125$ record; indeed they were the worst in the NFL. So, according to procedure, the Bills got first draft choice in 1985. They also got a new coach a quarter of the way through the season. Even so, the 1985 performance was another dismal $U = 0.125$.

In 1986, the Bills got a new quarterback and another new coach; they finished with a more respectable $U = 0.250$. In 1987, the team's performance rose to $U = 0.467$. In 1988, with a $U = 0.750$ record, the Bills wrapped up the division race and won the playoffs. After a 1989 slump to $U = 0.563$, the team rebounded to a very commendable performance of $U = 0.813$ during the 1990 and 1991 seasons and $U = 0.688$ during 1992.

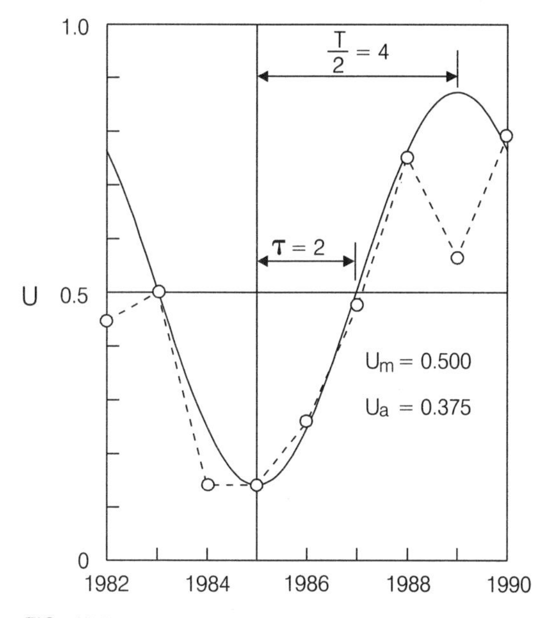

FIG. 12.5

Buffalo Bills near the year 1985. Comparison of actual performance with computed performance. Period $T = 8$ yr and delay time $\tau = 2$ yr.

This information about the Bills' performance from 1982 to 1990 is displayed in figure 12.5. The small open circles with the dashed lines show the actual record. The solid line is a plot of equation (12.14) with $U_m = 0.500$, $U_a = 0.125$ (point matched at 1985), and $T = 8$. The turnaround or delay time, τ, is shown in the figure. Although the oscillatory pattern of performance of the Buffalo Bills, as seen in figures 12.2 and 12.5, is especially clear, it is noteworthy that, without exception, all twenty-eight teams of the NFL display more or less similar patterns of peaks and troughs.

A Final Word

We have seen that our simple mathematical model, featuring a discrete-delay equation, can be applied—realistically, we hope—in the analysis of performance of a well-organized system, NFL professional football. We have used football as a vehicle to look at an interesting and important area of mathematics: discrete- and distributed-delay phenomena.

This kind of mathematics comes up in a great many fields of science and engineering: aeronautics, biology, control theory, dynamics, ecology, economics, epidemiology, physiology, and medicine. The NFL performance cycles have analogues in biological cell cycles, business cycles, climate cycles, construction cycles, insect and wildlife cycles, trade cycles, and so on.

Appendix: For Those Who Want to Do More with Statistics

If you are looking for a suitable subject for a technical report in your course in statistics or sports science, you may want to select one or more of the following topics.

TOPIC 1. For the thirty-three-year period from 1960 through 1992, which NFL teams have the largest and smallest average values $\overline{U}$, and how are the twenty-eight values distributed?

Response. This one is quite straightforward. The answers: (*a*) largest $\overline{U}$ is the 0.625 of the Los Angeles Raiders and (*b*) smallest $\overline{U}$ is the 0.298 of the Tampa Bay Buccaneers. The San Diego Chargers ($\overline{U} = 0.504$) and the Indianapolis Colts ($\overline{U} = 0.499$) are in the middle. The histogram of the twenty-eight $\overline{U}$ values is shown in figure 12.6.

As we saw in earlier chapters, the *normal* distribution is

$$N = \frac{N_0}{\sigma\sqrt{2\pi}} e^{-(\overline{U} - U_*)^2/2\sigma^2}, \tag{12.16}$$

where N is the number of teams having average value $\overline{U}$. In this equation, N_* is the mean and σ is the standard deviation of the twenty-eight $\overline{U}$ values. Since seven of the twenty-eight teams played for periods less than the thirty-three years, the mean value, $U_* = 0.495$, is a bit smaller than $U_m = 0.500$. Also, $\sigma = 0.0685$ and $N_0 = 28$. The computed normal distribution, equation (12.16), shown as the solid line in figure 12.6, fits the data fairly well.

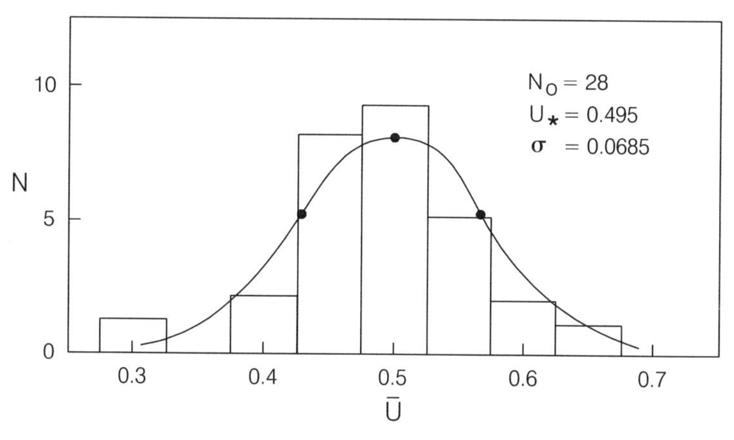

FIG. 12.6

Histogram of the average annual performance, $\bar{U}$, of the twenty-eight teams of the NFL, 1960 through 1992

TOPIC 2. Over the thirty-three-year period, there were 864 "team seasons," ranging from 21 during the 1960 season to 28 during the 1976 and subsequent seasons. How are these 864 U values distributed numerically?

Response. The histogram of the U values is presented in figure 12.7. A statistical analysis gives $U_m = 0.500$, of course, and $\sigma = 0.207$. The histogram has a very interesting shape: it is apparently symmetrical about U_m and it is distinctly *bimodal* (i.e., it exhibits *two* maxima). Two questions: Why should it be bimodal and what is an equation for the probability density distribution that fits the histogram shown in figure 12.7?

TOPIC 3. Again, we know that $U_m = 0.500$ for each of the thirty-three years. But what about the standard deviation, σ, (i.e., the "spread") of the U values? Was it about the same for each season or did it vary from year to year?

Response. For each of the thirty-three years we compute the average value U_m of the annual performances (0.500, naturally) and the standard deviation σ. Then, as shown in figure 12.8, we plot σ versus time. Very interesting. This plot seems to exhibit the same kinds of peaks and troughs we see in figures 12.2 and 12.3. The average

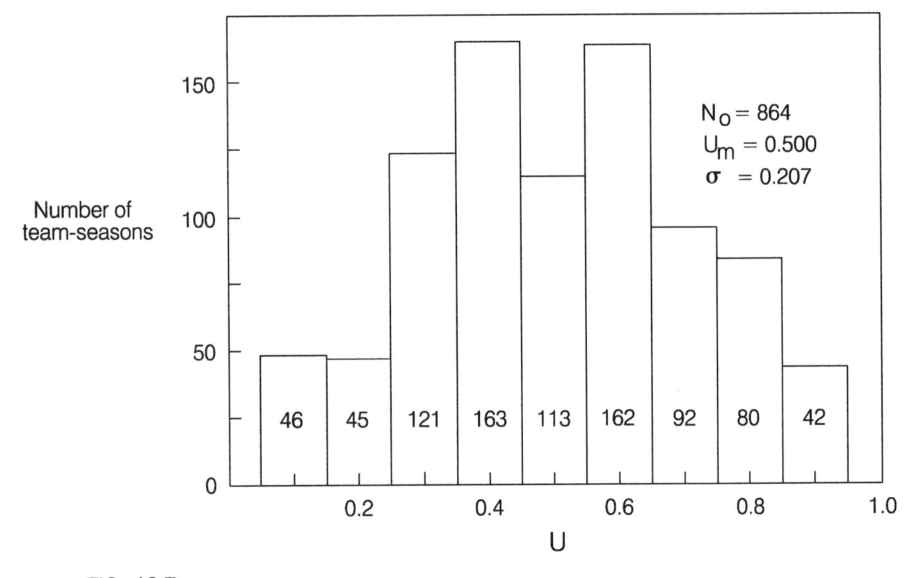

FIG. 12.7

Histogram of the number of team-seasons with the indicated values of U in the NFL, 1960 through 1992

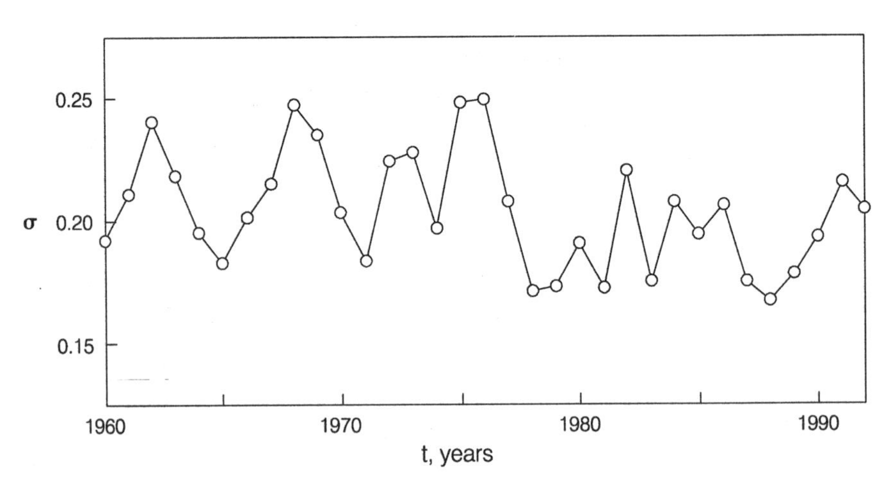

FIG. 12.8

Plot of the standard deviations, σ, of the annual performances, U, of NFL teams, 1960 through 1992

period is about $\bar{T} = 7.5$ yrs and the average standard deviation is $\bar{\sigma} = 0.202$. You might want to do the histogram on this.

In any event, here are some good questions. Why is there a rather distinct trend of seasons with "large" extremes of League-wide values of U (i.e., large σ) followed by seasons with smaller extremes (small σ)? Why is the apparent period about seven or eight years? Could this somehow be due to the effects of the two-year "delay time"? You might want to look into this matter. It may be helpful to utilize the relationships presented in the following paragraph.

The mean value m and variance σ^2 of a continuous function $y = f(x)$ are given, respectively, by the equations

$$m = \int x f(x)\, dx \qquad \text{and} \qquad \sigma^2 = \int (x - m)^2 f(x)\, dx. \qquad (12.17)$$

Using equation (12.14) to define $U(t)$, try to get an expression $\sigma = \sigma(t)$. Recall that, by definition, the standard deviation is the square root of the variance.

TOPIC 4. This is an appropriate topic to conclude our several chapters dealing with statistics and probability. There are a lot of commercials during football games, right? About fifteen minutes every hour. How many television channels need to be broadcasting football games simultaneously to reasonably assure that we can avoid commercials by switching channels, and, by doing so, continue to watch football?

Response. This problem is somewhat similar to the "chimpanzees on typewriters" problem we had in chapter 11. We let p (single trial "success") $= 45/60 = 3/4$ and q (single trial "failure") $= 15/60 = 1/4$. Proceeding as before, we determine that with one channel, the probability is 75% we can get football, with two channels 93.8%, with three channels 98.4%, and with four 99.6%.

The drawback, of course, is that this procedure does give you sore fingers and a confused brain.

13

A Tower, a Bridge, and a Beautiful Arch

Since the dawn of civilization, mathematics has been utilized to provide information needed to design and build massive structures. The earliest example is the Great Pyramids of Egypt. There was substantial knowledge of geometry and trigonometry at the time these spectacular feats of engineering were carried out (2650 to 2500 B.C.). The Pyramids are still there, of course, and are in remarkably good shape considering their forty-five centuries of exposure to the ravages of nature and mankind.

The civil and military engineers of Rome constructed massive buildings, bridges, aqueducts, and forts throughout the empire. Architects and engineers of the Middle Ages knew the principles of mechanics and structural stability necessary to design and construct the enormous beautiful cathedrals of Europe. These remarkable engineering achievements were feasible because of the mathematical and technological knowledge possessed by their designers and builders in those earlier days.

We are now going to examine three famous structures of more recent times. The first is the Eiffel Tower in Paris; it was completed in 1889. The second is the Golden Gate Bridge in San Francisco, which was opened to traffic in 1937. The third is the Gateway Arch in Saint Louis, finished in 1965.

As you anticipate, we are going to look into some of the mathematics involved in these very impressive structures. As we shall see, the main columns of the Eiffel Tower are described by a *logarithmic* equation, the shape of the cables of the Golden Gate Bridge is a *parabolic* curve, and the profile of the Gateway Arch is an inverted *catenary*.

The Eiffel Tower in Paris

If there is a "most famous structure in the world" it must be the beautiful Eiffel Tower. It was constructed for the Universal Exhibition of 1889 to commemorate the one hundredth anniversary of the French Revolution.

The tower, shown in figure 13.1, was designed by Gustave Eiffel (1832–1923), one of the greatest engineers of France. He was born in Dijon and studied engineering at the Ecole Centrale in Paris. Before he built the tower, he had constructed a great many large bridges and buildings in France and numerous other countries.

Indeed, Eiffel had also designed the framework of the Statue of Liberty, a gift to America from France completed in 1886. Years later, after the tower had been completed, Eiffel carried out studies on the aerodynamics of wind forces; not surprisingly, he utilized the tower as his research facility. The story of the life of Eiffel and a description of his many engineering achievements are given by Loyrette (1985).

The height of the tower is 984 feet and its square base has an overall side length of 414 feet. The structure is made of wrought iron and has a total weight of 8,090 tons. There are 15,000 separate pieces and 2,5000,000 rivets in the structure. The time of construction was slightly over two years.

Structural Analysis of the Eiffel Tower

We are now going to analyze the main structural components of the Eiffel Tower, as we shall also do for the Golden Gate Bridge and the Gateway Arch. Here is our plan. First, we are going to pretend that the actual iron framework is replaced by an equivalent or virtual material, such as light-weight plastic, that

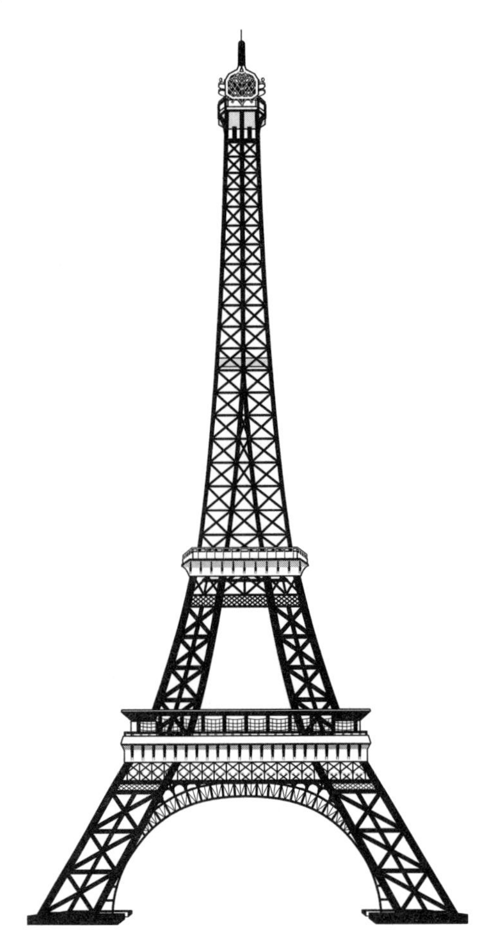

FIG. 13.1

The Eiffel Tower. During its construction and for a while afterward there was strong opposition to the unsightly structure. Fortunately, the Parisians changed their minds. (Photograph provided by French Cultural and Scientific Services.)

has the same total weight and occupies the same overall volume as the framework. Second, we shall slice a thin horizontal section from the tower to obtain a so-called free-body diagram. This is shown in figure 13.2.

The small segment illustrated is in vertical equilibrium; otherwise, according to Newton's second law of motion, it would move. So the downward forces are equal to the upward forces. The

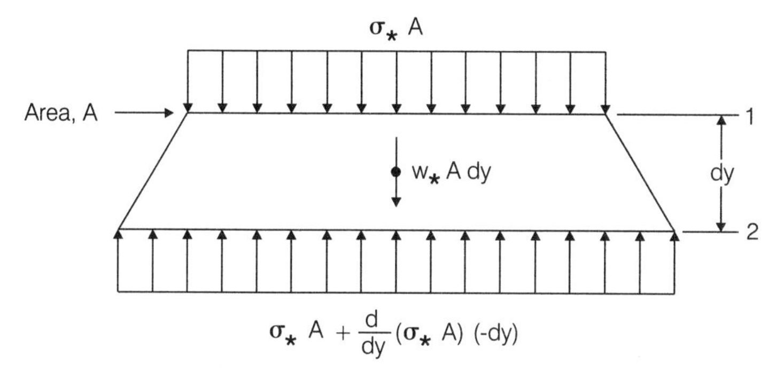

FIG. 13.2

Free-body diagram for vertical static equilibrium

downward force, acting on plane 1, is the compression force, $\sigma_* A$, where σ_* is the compressive stress (lb/ft^2) and A is the cross-sectional area (ft^2). In addition, there is the weight of the small elemental volume, $w_* A\,dy$, in which w_* (lb/ft^3) is the specific weight of the material and dy is the thickness of the slice.

As usual, the y-axis is positive in the upward direction. So the upward compression force acting on plane 2, located a distance dy below plane 1, is $\sigma_* A + (d/dy)(\sigma_* A)(-dy)$. Equating the downward and upward forces, we obtain the following equation for equilibrium in the vertical direction:

$$w_* A = -\frac{d}{dy}(\sigma_* A). \tag{13.1}$$

For reasons of economy and safety, it is necessary to keep the compressive stress, σ_*, constant. As we shall see, this is why the tower bends outward as the elevation, y, decreases. Also, the area of the square cross section of the tower is $A = (2x)^2$. Some algebra reduces equation (13.1) to the form

$$\int_0^y dy = -\frac{2\sigma_*}{w_*}\int_{x_0}^x \frac{1}{x}\,dx. \tag{13.2}$$

The lower limits of the integrals indicate that when $y = 0$ (on the ground) then $x = x_0$ (half of the length of the base; $x_0 = 207$ ft).

Integration of equation (13.2) gives

$$y = -y_* \log_e \frac{x}{x_0}, \tag{13.3}$$

in which $y_* = 2\sigma_*/w_*$. Thus, a logarithmic equation describes the profile of the Eiffel Tower.

Next we measure the (x, y) coordinates of the profile at, say, twenty or twenty-five elevations of the tower. A much enlarged copy of figure 13.1 was utilized for this purpose. Substituting these (x, y) values in equation (13.3), we easily compute the height constant, y_*; the average value is $y_* = 301$ ft.

The total weight of the tower is given by the relationship

$$W = w_* \int_0^H A\,dy = 4w_* \int_0^H x^2\,dy. \tag{13.4}$$

From equation (13.3) we obtain $x = x_0 e^{-y/y_*}$. Substituting this into equation (13.4) and integrating gives

$$W = 4\sigma_* x_0^2 (1 - e^{-2H/y_*}). \tag{13.5}$$

Knowing that $W = 8{,}090$ tons $= 16{,}180{,}000$ lb, $x_0 = 207$ ft, $H = 984$ ft, and $y_* = 301$ ft, equation (13.5) gives $\sigma_* = 94.5\ \text{lb/ft}^2$.

Finally, since $y_* = 2\sigma_*/w_*$ and since $y_* = 301$ ft and $\sigma_* = 94.5\ \text{lb/ft}^2$, we easily determine that $w_* = 0.628\ \text{lb/ft}^3$. Don't forget that our tower is still completely filled with the equivalent very-light-weight material.

Now we shift back to our wrought-iron structure. Iron has a specific gravity of 7.4; this means it weighs 7.4 times more than water. Since $w(\text{water}) = 62.4\ \text{lb/ft}^3$, then the actual specific weight is $w(\text{iron}) = 462\ \text{lb/ft}^3$.

The ratio of the specific weight of the equivalent material, w_*, to the specific weight of the actual iron, w, is given by $w_*/w = 0.628/462 = 0.00136$. We shall assume that the compressive stresses in the two materials follow this same ratio. That is, $\sigma_*/\sigma = 0.00136$. Accordingly, the compressive stress in the iron is $\sigma = 94.5/0.00136 = 69{,}490\ \text{lb/ft}^2 = 482\ \text{lb/in}^2$. Of course, this value is simply the average stress in the iron; maximum values would be considerably more. Even so, nowadays engineers utilize design stresses considerably larger than these, even for wrought iron. Eiffel was wisely conservative.

We now have everything we need. Here is a summary:

height, $H = 984$ ft
Length, $L = 2x_0 = 414$ ft
height constant, $y_* = 2\sigma_*/w_* = 2\sigma/w = 301$ ft

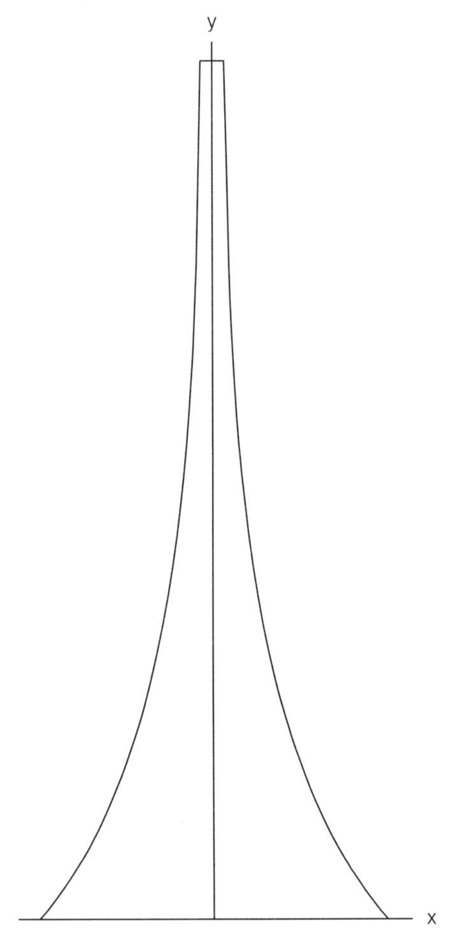

FIG. 13.3

The computed profile of the Eiffel Tower. The equation of the profile is $y = -y_* \log_e x/x_0$, where y is the elevation and $2x$ is the width.

actual compressive stress, $\sigma = 69{,}490\ \text{lb/ft}^2$
virtual compressive stress $= 94.5\ \text{lb/ft}^2$
actual specific weight $w = 462\ \text{lb/ft}^3$
virtual specific weight $= 0.628\ \text{lb/ft}^3$

Using equation (13.3), we calculate and plot the profile of the Eiffel Tower. The result is shown in figure 13.3. Note that the computed and actual profiles look almost the same. In the actual tower, the profile is a straight line from the base ($y = 0$) to the lowest observation deck ($y = 175$). It is pointed out that the height constant, y_*, is not exactly a constant $y_* = 301$. In the computations it varied from $y_* = 275$ at $y = 175$ ft to $y_* = 345$ at $y = 850$ ft. Our assumption of constant y_* produces a computed profile that is a bit wider than the actual tower for the smaller elevations and slightly narrower for the larger.

Eiffel Tower Problems

PROBLEM 1. Show that the volume of the tower is given by

$$V = 2x_0^2 y_* (1 - e^{-2H/y_*}).$$

PROBLEM 2. In figures 13.1 and 13.3, the profile of the tower is in the (x, y) plane. Using equation (13.3), show that the computed slope at the base is $\theta_0 = 55.5°$. In the diagonal (r, y) plane, where $r = \sqrt{2}x$, show that the calculated slope at the base is $\phi_0 = 45.8°$. What are the slopes at the top of the tower?

PROBLEM 3. The lengths of each of the columns defining the profile at the four corners of the tower are about $S = 1{,}050$ ft. Confirm this by using the basic equation for determining arc lengths:

$$S = \int \sqrt{dr^2 + dy^2}.$$

PROBLEM 4. This problem is concerned with the compression force at the bottom of the tower. At each of the four corners there is a square base measuring 87 feet on each side. Dividing the weight of the tower

FIG. 13.4

The Golden Gate Bridge. (Photograph provided by Golden Gate Bridge, Highway and Transportation District.)

by the total base area gives a compression load of 535 lb / ft^2. For comparison, the base load is 21,500 lb / ft^2 for the towers of the Golden Gate Bridge and 18,800 lb / ft^2 for the Gateway Arch.

PROBLEM 5. Confirm that the distance D to the horizon from a point at an elevation H is given by the very simple formula D (miles) = $1.2\sqrt{H \text{ (ft)}}$. Thus, from the top of the 984-foot Eiffel Tower, the distance to the horizon is about 38 miles.

The Golden Gate Bridge in San Francisco

Although it no longer holds first place, in 1937 when it was completed, the Golden Gate Bridge was the longest suspension bridge in the world. By "longest" we refer to the length L of the cable-supported center span between the two towers. For the Golden Gate Bridge, $L = 4{,}200$ ft.

The bridge is shown in figure 13.4. Including the two side spans, each 1,125 feet long, the total length is 6,450 feet. The towers are 746 feet in height.

The chief engineer of the enormous undertaking was Joseph Strauss (1870–1938). Born in Ohio, he studied mathematics and mechanics at the University of Cincinnati. Later, he established an engineering company and built many bridges throughout the world. Though small in size, Strauss was a tremendous organizer and promoter. Without his persistence, the Golden Gate Bridge would not have happened until years later. The story of the design and construction of the bridge, and many vignettes about the people involved, are given in the book by Van der Zee (1986).

The towers of this spectacular structure are supported by massive piers resting on solid rock. The two 36-inch diameter steel cables are secured to 60,000-ton reinforced concrete anchorages at each end of the bridge. The roadway section is 90 feet wide and is supported by steel trusses 25 feet deep, connected every 50 feet by suspenders attached to the cables.

The design load for the truss-supported roadway sections was 19,000 pounds per foot of bridge length. Of this, 15,000 pounds was dead load and 4,000 pounds live load. In addition, the weight of the cables was 6,000 pounds. So the total design load was 25,000 pounds per foot of bridge length.

It is noted that the cables represented 24% of the total load and the truss sections 76%. As we shall see, this distinction between cable load and truss load is important in our mathematical analysis.

As in our examination of the Eiffel Tower, we construct a free-body diagram. This is done by cutting the bridge and cable at the midpoint of the center span, $x = 0$, and at some other point a distance x to the right. This gives us the free-body diagram shown in figure 13.5. Again, the section is in equilibrium. This means that the forces acting in the x direction balance each other and the forces in the y direction also balance each other.

From the above diagram we obtain the following results:

$$T_0 = T\cos\theta \qquad \text{and} \qquad W = T\sin\theta, \tag{13.6}$$

in which T_0 is the tension force in the cable at $x = 0$, T is the tension force at any value x, and W is the weight of the truss and

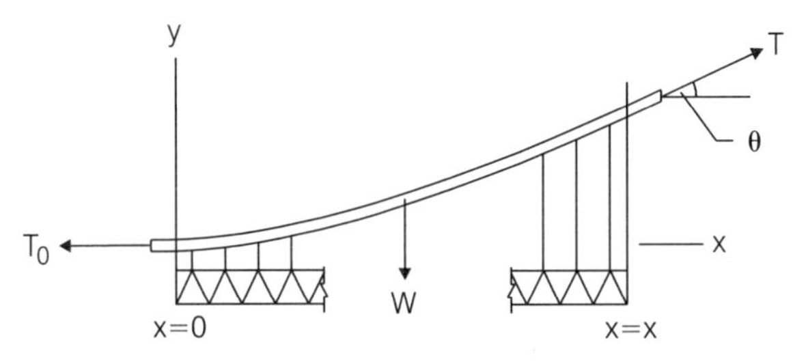

FIG. 13.5

Free-body diagram for a suspension bridge

cable between $x = 0$ and $x = x$. Dividing the second relationship in equation (13.6) by the first, we obtain

$$\frac{T \sin \theta}{T \cos \theta} = \tan \theta = \frac{W}{T_0}. \tag{13.7}$$

Now T_0 is a constant and $W = \gamma x$, where γ is the weight of the truss and cable per foot of bridge length. If the weight due to the truss is much larger than the weight due to the cable, then we can assume that γ is the same for all values of x. We mention this point because in the analysis of the Gateway Arch things are a bit different.

Now, since $\tan \theta = dy/dx$,

$$\frac{dy}{dx} = \frac{\gamma}{T_0} x, \tag{13.8}$$

which is easily integrated to give

$$y = \frac{\gamma}{2T_0} x^2. \tag{13.9}$$

This is the equation that describes the shape of the cable. Being ever alert, you have already noticed that this is a parabola. This is the graceful curve you see in figure 13.4.

Next, substituting $x = L/2$ and $y = H$ into equation (13.9) gives

$$T_0 = \frac{\gamma L^2}{8H}. \tag{13.10}$$

From this relationship, the tension force in the cable at midspan can be computed. This is the minimum tension force. If equation (13.9) is differentiated, we obtain

$$\frac{dy}{dx} = \tan\theta = \frac{\gamma}{T_0}x. \tag{13.11}$$

Again substituting $x = L/2$, we get

$$\tan\theta_0 = \frac{\gamma L}{2T_0}, \tag{13.12}$$

and from this equation the angle of the cable at the top of the tower can be calculated. Then from equation (13.6) the maximum tension force is computed.

Our mathematical analysis has produced quite a few interesting equations. Now let us substitute some numbers and compute some typical engineering results.

A Numerical Example

What is the tension force in the cables of the Golden Gate Bridge at the center of the main span and at the top of the Towers? We know that γ (design load per cable) = 12,500 lb/ft; L = 4,200 ft; tower height = 746 ft; bridge height = 276 ft; and H = 470 ft. So

$$T_0 = \frac{(12{,}500)(4{,}200)^2}{8(470)} = 58.7 \times 10^6 \text{ lb}.$$

This is the minimum tension force in each cable.

In addition, we determine that $\theta_0 = 24.1°$. So from equation (13.6) we ascertain that $T_m = 64.3 \times 10^6$ lb. This is the maximum tension force in the cable. Knowing the tension force and also the allowable stress for steel, engineers can calculate the necessary cross-sectional area and diameter of the cables.

This example is a simplified illustration of the type of computations structural engineers must carry out in the design of suspension bridges. Such computations are much more complicated when problems of wind loads, earthquake loads, temperature stresses, aerodynamic stability, and other factors are taken into account.

We conclude our example with the following. The total length of each of the two cables of the Golden Gate Bridge is 7,650 feet. Each cable is composed of 61 strands each containing 452 wires (nearly 1/4-inch diameter). So the total length of wire is about 80,000 miles. This is sufficient to go around the world more than three times.

Finally, the diameter of each cable is slightly over 36 inches and its weight is approximately 3,000 pounds per foot. If we divide the maximum force in each cable, $T_m = 64.3 \times 10^6$ lb, by the cross-sectional area of the cable, $A = (\pi/4)(36)^2$, we determine that the maximum tension stress is about 63,000 lb/in^2.

The Gateway Arch in Saint Louis

In 1947 a competition was held for the design of a structure, to be built in Saint Louis, to symbolize the opening of America's West following the Louisiana Purchase of 1803. The winner of this contest for the "Gateway to the West" was the noted young architect Eero Saarinen.

In the years that followed, Eero Saarinen and Associates carried out extensive analyses and made detailed plans of the structure now known as the Gateway Arch. Tragically, Saarinen died in 1961. However, his successors continued the work, construction was begun in early 1963, and in late 1965 the arch was completed.

The Gateway Arch is the main component of the Jefferson National Expansion Memorial located on the west bank of the Mississippi River in downtown Saint Louis. The memorial is

FIG. 13.6

Gateway Arch in Saint Louis. (Photograph provided by National Park Service.)

administered by the National Park Service. This beautiful and impressive structure is seen in figure 13.6. The base of the arch is situated in the middle of a large open plaza surrounded by many acres of trees and shrubs.

Eero Saarinen (1910–61) was born in Finland and came to the United States with his parents in 1923. He received his degree in architecture from Yale University in 1934. For several years he worked with his father, also an architect, on a number of projects. Later, he established Eero Saarinen and Associates; they designed many of America's most impressive buildings. An excellent reference describing these structures including his most noteworthy achievement—the Gateway Arch—is the book edited by his wife, Saarinen (1962).

The view of the Gateway Arch shown in figure 13.6 is looking to the east. The river is just beyond, and downtown St. Louis is behind you. The structure is 630 feet high and the base length is also 630 feet. The cross section of the arch is an equilateral triangle with side length tapering from 54 feet at the base to 17 feet at the top.

The profile of the graceful structure is described by a mathematical curve called the inverted catenary. This is the shape—upside down—of a deeply sagging flexible cable. Our assignment is to determine the equation of the curve.

Our analysis begins with the free-body diagram shown in figure 13.7. It is similar to the diagram of figure 13.5 for the Golden Gate Bridge, with one major difference. In the bridge problem, γ is the weight per unit length of *bridge*; accordingly, the total weight from $x = 0$ to $x = x$ is $W = \gamma x$. In contrast, this time there is no truss section involved. We simply have a sagging flexible cable with γ being the weight per unit length of *cable*. Consequently, the total weight from $x = 0$ to $x = x$ is $W = \gamma s$, where s is the arc length of cable.

As before, we obtain the relationships for static equilibrium given by equations (13.6) and (13.7). However, this time $W = \gamma s$ instead of $W = \gamma x$.

By the way, $\gamma = \rho g$, where γ is the *weight* per unit length of cable, ρ is the *mass* per unit length, and g is the force per unit mass due to gravity ($g = 32.2\ \mathrm{ft}/s^2 = 9.82\ \mathrm{m}/s^2$). This is mentioned because in the "jumping rope" problem of the next

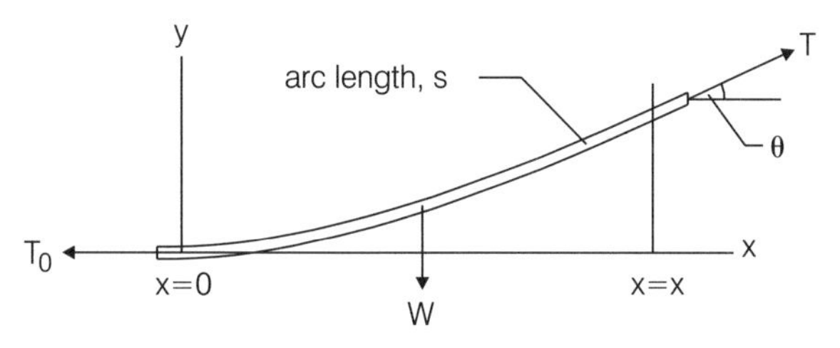

FIG. 13.7

Free-body diagram for a catenary

chapter we are going to replace gravitational force with centrifugal force.

Now, as we have indicated previously, the arc length of a curve is given by the equation

$$s = \int_0^s \sqrt{dx^2 + dy^2} = \int_0^x \sqrt{1 + \left(\frac{dy}{dx}\right)^2}\, dx. \tag{13.13}$$

Accordingly, from equation (13.7), with $\tan\theta = dy/dx$, we have

$$\frac{dy}{dx} = \frac{\gamma}{T_0} s = \frac{\gamma}{T_0} \int_0^x \sqrt{1 + \left(\frac{dy}{dx}\right)^2}\, dx. \tag{13.14}$$

The easiest way to solve this equation is to let $p = dy/dx$ and then differentiate. This gives

$$\frac{dp}{dx} = \frac{\gamma}{T_0}\sqrt{1 + p^2}. \tag{13.15}$$

We impose the condition that when $x = 0$, then $y = 0$ and $p = 0$. Integrating this equation, we obtain the answer

$$y = \frac{T_0}{\gamma}\left(\cosh\frac{\gamma x}{T_0} - 1\right), \tag{13.16}$$

where "cosh" is the hyperbolic cosine. This is the equation of the catenary curve. Incidentally, for small values of $\gamma x/T_0$, and using the series approximation $\cosh z = 1 + z^2/2$, equation (13.16) reduces to equation (13.9). In essence, this means that if the sag of the cable is small, the catenary becomes a parabola.

As before, we substitute $x = L/2$ and $y = H$ into equation (13.16) to get

$$H = \frac{T_0}{\gamma}\left(\cosh\frac{\gamma L}{2T_0} - 1\right). \tag{13.17}$$

We use this equation to compute the minimum tension force, T_0.

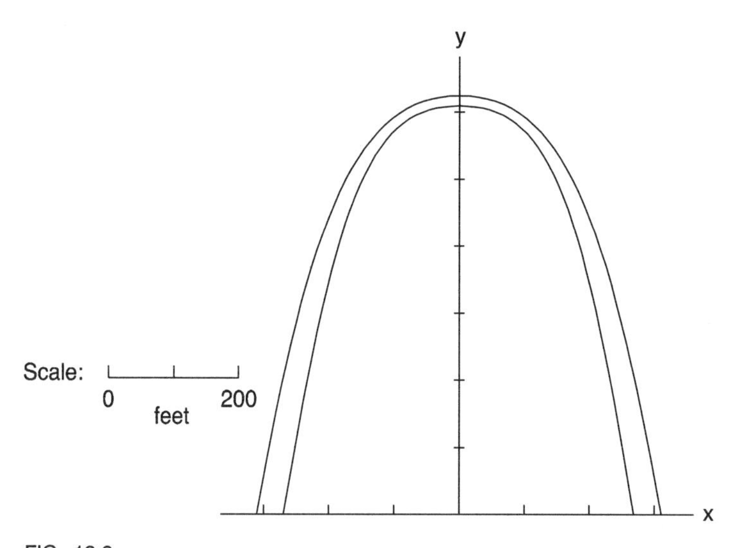

FIG. 13.8

The computed profile of the Gateway Arch

Finally, using equation (13.13), we determine the total arc length of the catenary between the two supports. The answer is

$$S = \frac{2T_0}{\gamma} \sinh \frac{\gamma L}{2T_0}, \tag{13.18}$$

in which "sinh" is the hyperbolic sine. Scientific calculators, of course, have keys for the trigonometric functions (sin, cos, tan). In addition, nearly all of them have keys for the hyperbolic functions (sinh, cosh, tanh).

All this analysis has enabled us to construct mathematically the Gateway Arch shown in figure 13.8. Our next step is to "freeze" our flexible cable, turn it upside down, and shift the mathematical origin. We also change the cross section from a circle to an equilateral triangle and taper the side length of the triangle from b_0 at $s = 0$ to B at $s = S/2$. With the new origin, the equation of the *outer* profile of the arch is

$$y = H - a\left(\cosh \frac{x}{a} - 1\right), \tag{13.19}$$

where a replaces T_0/γ. Letting $y = 0$ when $x = L/2$, we obtain

$$H = a\left(\cosh \frac{L}{2a} - 1\right). \tag{13.20}$$

For the Gateway Arch, $H = 630$ ft and $L = 630$ ft. Substituting these numbers in equation (13.20) yields the value $a = 127.7$ ft.

A similar analysis is carried out for the *inner* profile,

$$y = H_* - c\left(\cosh \frac{x}{c} - 1\right), \tag{13.21}$$

where $H_* = H - (\sqrt{3}/2)b_0 = 630 - (\sqrt{3}/2)(17) = 615$ ft. This time, when $y = 0$, $x = 315 - 1(\sqrt{3}/2)(54) = 268$ ft. So we get $c = 101.4$ ft.

We now have the information we need to compute all kinds of things. For example, what is the angle of the outer profile at the base of the arch? First, the derivative of equation (13.19) is

$$\frac{dy}{dx} = \tan\theta = -\sinh \frac{x}{a}. \tag{13.22}$$

Substituting $x = 315$ ft and $a = 127.7$ ft into this equation gives $\theta_0 = 80.3°$.

Another example: What is the arc length of the outer profile? Utilizing equation (13.13), you should obtain the result $S =$ 1,494 ft.

And a third example: What is the surface area of the arch and what is its volume? The answers: $A = 155{,}000$ ft^2 and $V =$ 866,500 ft^3.

The Catenoid

We conclude our chapter with a few words about one of the remarkable features of the catenary. First, imagine that a soap film connects two parallel hoops of the same diameter held a certain distance apart. Can you visualize the shape of the soap film?

In any event, suppose that we also rotate a catenary curve, $y = a \cosh(x/a)$ about its so-called directrix, the straight line located a distance a below the low point of the curve. In doing so, we generate a body of revolution called the *catenoid.* Interestingly, the surface of this body has precisely the same shape as our soap film.

Using a branch of mathematics called the *calculus of variations,* it can be shown that the catenoid is the body of revolution giving minimum area to the soap film connecting the two parallel hoops. There is a quite remarkable book by Isenberg (1992) that deals with topics like these.

14

Jumping Ropes and Wind Turbines

In the previous chapter, in our study of the famous Gateway Arch in Saint Louis, we analyzed the mathematical curve called the catenary. You will remember that this is the curve that describes the shape of a sagging flexible cable. We obtained the following equation for the catenary:

$$y = H\left[1 - \frac{\cosh(x/a) - 1}{\cosh(L/2a) - 1}\right], \tag{14.1}$$

in which cosh is the hyperbolic cosine and $a = T_0/\rho g$, where ρ is the mass of the cable per unit length, g is the force per unit mass due to gravity ($g = 32.2\ \text{ft/s}^2 = 9.82\ \text{m/s}^2$), and T_0 is the tension force in the cable at the lowest point. This equation, shown in figure 14.1, says that when $x = 0$, $y = \text{H}$ and when $x = L/2$, $y = 0$. You will also remember that we then froze this cable, turned it upside down, substituted $H = 630$ ft, $L = 630$ ft, and $a = 127.7$ ft, and thus created the profile of the beautiful structure called the Gateway Arch.

Now, instead of a cable or rope that sags due to the force of gravity, we are going to analyze a *jumping rope*—like children use to have fun and adults use to stay in shape. As we shall see, in this case the force that determines the shape of the rope is no longer the gravity force—it is *centrifugal* force. As we shall also

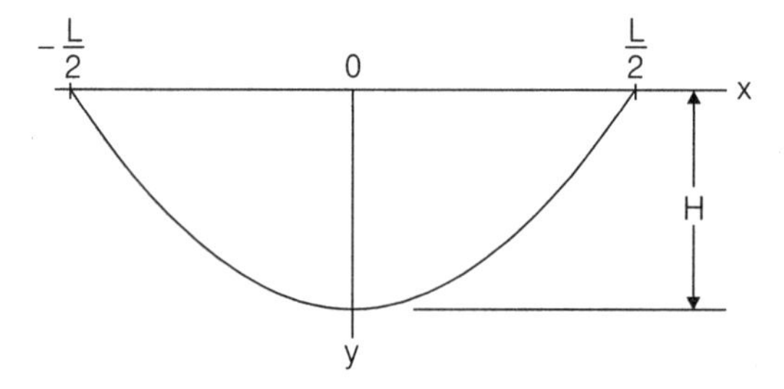

FIG. 14.1

Definition sketch for the catenary

see, we now have a somewhat more difficult problem. However, before we get started on that, let us look at a couple of other problems.

Periods of Pendulums and Lengths of Ellipses

Periods of Pendulums

In our problem on the Gateway Arch, we pretend we have a weight connected to a long rope hooked to the top of the arch and extending almost to the ground. This gives a pendulum of length $L = 630$ ft. As we know, the period for one complete oscillation of this long pendulum is given by the equation

$$T = 2\pi\sqrt{\frac{L}{g}}\,. \qquad (14.2)$$

Substitution of $L = 630$ ft and $g = 32.2\ \text{ft/s}^2$ gives $T = 27.8$ s.

We need to mention that equation (14.2) is valid only when the deflection angle of the pendulum, θ_0, is "small"—say about 5° or so from the vertical. Now we suppose that the initial deflection angle is not small. What is the oscillation period if, for example, $\theta_0 = 45°$ or more? This is a pretty hard problem. It turns out that the answer is

$$T = 4\sqrt{\frac{L}{g}}\,K(k), \qquad (14.3)$$

where $k = \sin(\theta_0/2)$. The quantity $K(k)$ has the definition

$$K(k) = \int_0^{\pi/2} \frac{d\phi}{\sqrt{1 - k^2 \sin^2 \phi}}. \tag{14.4}$$

This is known as the *complete elliptic integral of the first kind*. Even though this definite integral is nowhere near as familiar as, say, the trigonometric or hyperbolic functions, it is a well-tabulated quantity and is easy to utilize.

To convince you, here is a numerical example. Suppose that somehow the weighted end of our Gateway Arch pendulum is swung out and raised to a height of 630 feet, that is, $\theta_0 = 90°$, and then released. Accordingly, $k = \sin(90°/2) = 1/\sqrt{2} = 0.7071$. Turning to a table of elliptic integrals, we obtain $K(k) = 1.8541$. Substituting this and the other numbers into equation (14.3) gives $T = 32.8$ s.

The important thing is that the period T for this large-amplitude oscillation is about 5.0 seconds longer than the period for the small-amplitude oscillation. Amazing. Why not try this experiment the next time you use the park or school yard swings?

Lengths of Ellipses

The area of a circle is $A = \pi R^2$, where R is the radius. The area of an ellipse is $A = \pi ab$, where a is half the length of the major axis of the ellipse and b is half the length of the minor axis. The circumference or arc length of a circle is $S = 2\pi R$. The arc length of an ellipse is . . . what?

Well, it turns out that even though the area of an ellipse is easy to determine, its circumference or arc length is anything but easy. Indeed, until the early part of the nineteenth century, this was an unsolved problem in mathematics. Then some of the great mathematicians, intrigued by ellipses, developed what we now call elliptic functions and elliptic integrals. The two who made the most significant early contributions on these subjects were the French mathematician Adrien-Marie Legendre (1752–1833) and the German mathematician Carl Gustav Jacob Jacobi (1804–51).

The answer to the question of the arc length of an ellipse is

$$S = 4aE(k), \tag{14.5}$$

in which $k = \sqrt{1 - (b/a)^2}$ and

$$E(k) = \int_0^{\pi/2} \sqrt{1 - k^2 \sin^2 \phi}\, d\phi. \tag{14.6}$$

This quantity is known as the *complete elliptic integral of the second kind*. This definite integral is also a well-tabulated function. Tables of elliptic integrals, and a great many other mathematical functions, are given in Abramowitz and Stegun (1965).

Here is a numerical example. As is well known, the earth is an oblate ellipsoid. Thus, the equatorial radius of the earth is $a = 6{,}378.4$ km and the polar radius is $b = 6{,}356.9$ km. How much further is it to go around the world on the equator than is it to go through the North and South poles?

Well, to go along the equator is like going around a circle; the distance is obviously

$$S_e = 2\pi a = 40{,}076.7 \text{ km}.$$

However, to go through the poles, we have to go around an ellipse. The value of k is $k = \sqrt{1 - (b/a)^2} = 0.0820$. Turning to our tables we obtain $E(k) = 1.5682$. Substituting into equation (14.5) gives

$$S_p = 4aE(k) = 40{,}010.4 \text{ km}.$$

So we get the answer: $\Delta S = S_e - S_p = 66.3$ km. You might want to bear this in mind when you plan your next trip around the world. The polar route is a bit shorter.

Jumping Ropes or, to Be Scientific, Troposkeins

After that brief excursion through pendulums and ellipses, we return to the problem of the jumping rope or troposkein (Greek for a "turning rope"). The definition sketch for the problem is seen in figure 14.2. A rope of constant mass per unit length, ρ,

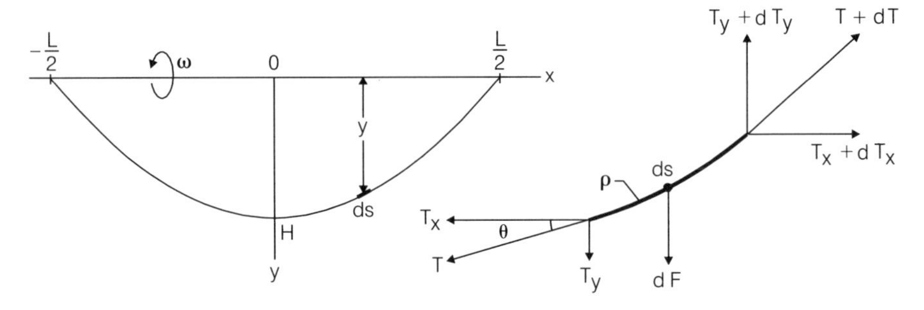

FIG. 14.2

Definition sketch for the jumping rope or troposkein

rotates about the x-axis at an angular velocity ω radians per second. (Note: 1 radian $= 360°/2\pi = 57.296°$.) We ignore the effect of gravity. The distance between the rope connections on the x-axis is L; the maximum displacement is H. As in the catenary problem, we want to determine the equation of the curve, $y = f(x)$.

The various forces acting on a very short section of the rope, ds, are shown in the figure. As in the catenary analysis, the tension force in the x direction is constant and equal to T_0, the value at $x = 0$.

The centrifugal force, dF, is balanced by the tension force in the y direction, dT_y. Now recall that the centrifugal force acting on any object moving along a curved path is $F = mU^2/r$, in which m is the mass of the object, U is its velocity, and r is the radius of curvature of the path. In our problem, $U = \omega y$ and $r = y$. Consequently, $dF = (\rho\, ds)(\omega y)^2/y = \rho\omega^2 y\, ds$, and hence $dT_y = \rho\omega^2 y\, ds$.

The assumption is made that the effects of aerodynamic drag on the rope can be neglected.

With reference to the figure, we obtain

$$\tan\theta = \frac{T_y}{T_x} = -\frac{dy}{dx}, \tag{14.7}$$

and, differentiating,

$$\frac{d^2y}{dx^2} = -\frac{d}{dx}\left(\frac{T_y}{T_x}\right) = -\frac{1}{T_0}\frac{dT_y}{dx}, \tag{14.8}$$

which becomes

$$\frac{d^2y}{dx^2} = -\frac{1}{T_0}\rho\omega^2 y\frac{ds}{dx} = -\frac{\rho\omega^2}{T_0}y\sqrt{1+\left(\frac{dy}{dx}\right)^2}, \tag{14.9}$$

since $ds = \sqrt{dx^2 + dy^2}$. This is the differential equation that must be solved. The boundary conditions are (a) $x = 0$, $dy/dx = 0$ and (b) $x = L/2$, $y = 0$.

The easiest way to proceed is to let $p = dy/dx$. Consequently, $d^2y/dx^2 = p(dp/dy)$, and equation (14.9) becomes

$$p\frac{dp}{dy} = -\frac{\rho\omega^2}{T_0}y\sqrt{1+p^2}. \tag{14.10}$$

Separating the variables and installing the integral signs gives

$$\int_0^p \frac{p\,dp}{\sqrt{1+p^2}} = -\frac{\rho\omega^2}{T_0}\int_H^y y\,dy. \tag{14.11}$$

The lower limits on the integrals provide boundary condition (a): when $x = 0$ then $y = H$ and $dy/dx = p = 0$.

Carrying out the indicated integration, we obtain the intermediate result

$$p = \sqrt{c(H^2 - y^2)[2 + c(H^2 - y^2)]}, \tag{14.12}$$

where $c = \rho\omega^2/2T_0$.

At this point, we let $y = H\sin\phi$. Substituting this relationship into equation (14.12) and carrying out some algebra, we set up the following integral, utilizing boundary condition (b):

$$\int_{L/2}^x dx = -\frac{k}{cH}\int_0^\phi \frac{d\phi}{\sqrt{1 - k^2\sin^2\phi}}, \tag{14.13}$$

in which

$$c = \frac{\rho\omega^2}{2T_0}; \qquad k^2 = \frac{1}{1 + 4T_0/\rho\omega^2 H^2}. \tag{14.14}$$

Integration of equation (14.13) and more algebra provides our final answer:

$$x = \frac{L}{2}\left[1 - \frac{F(k,\phi)}{K(k)}\right]; \qquad y = H\sin\phi, \tag{14.15}$$

where $K(k)$ is given by equation (14.4) and

$$F(k,\phi) = \int_0^{\phi} \frac{d\phi}{\sqrt{1 - k^2\sin^2\phi}} \tag{14.16}$$

is the (incomplete) elliptic integral of the first kind.

In this jumping rope—troposkein—problem, it is not possible to obtain a solution in explicit form, $y = f(x)$, as we were able to do in the case of the catenary. We have to settle for an answer in *parametric* form, $x = g(\phi)$ and $y = h(\phi)$. This is really not much of a hardship.

Another thing: It is fairly easy to apply equation (14.16). As in the case of the complete ($\phi = \pi/2$) elliptic integrals, expressed by equations (14.4) and (14.6), the quantity $F(k,\phi)$ is well tabulated.

Jumping Rope Problems

PROBLEM 1. Show that the ratio H/L and k are related by the expression

$$\frac{H}{L} = \frac{k}{(1-k^2)K(k)}. \tag{14.17}$$

PROBLEM 2. Show that the total length of the rope, S, is given by the equation

$$\frac{S}{L} = \frac{2}{(1-k^2)}\frac{E(k)}{K(k)} - 1. \tag{14.18}$$

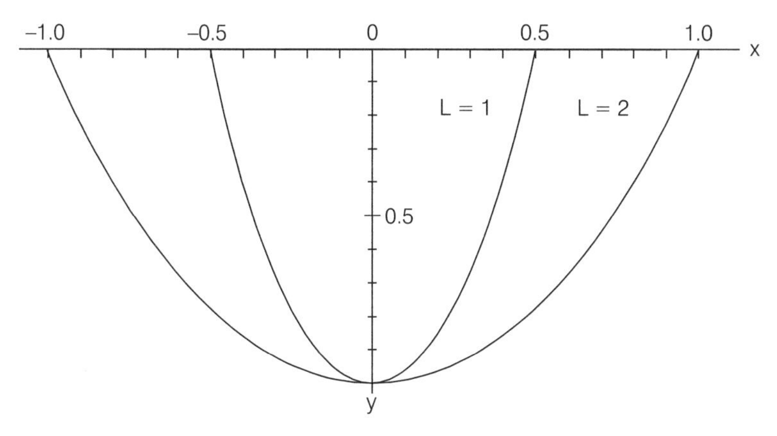

FIG. 14.3

Examples of jumping rope or troposkein curves: $H = 1.0$, $L = 1.0$ and $H = 1.0$, $L = 2.0$

PROBLEM 3. Suppose that we want the values $H = 1.0$ m and $L = 1.0$ m. Compute and plot the curve and determine its length.

An iteration of equation (14.17) gives $K = 0.776$. By the way, most tables of elliptic integrals list values in terms of $\alpha = \arcsin k$, instead of k. So we have $\alpha = 50.9°$, $K(k) = 1.9521$, and $E(k) = 1.2972$. From a table of elliptic integrals and equation (14.15), we obtain the curve seen in figure 14.3. Using equation (14.18), the length of the curve is $S = 2.34$ m.

Very interesting. Why not invent and solve your own example—perhaps with dimensions suitable for three kids jumping the same rope at the same time? Would $H = 1.25$ m and $L = 3.0$ m be about right?

Darrieus Vertical-Axis Wind Turbines

The troposkein curve has a very useful practical application in wind energy technology. The flexible jumping rope we just analyzed is in pure tension. There is no compression force or bending force. So if we make a wind turbine blade, of aluminum or stainless steel, in the shape of a troposkein curve, and operate it at design speed, there will be no risk of blade failure due to bending forces.

This is one of the reasons that led G. J. M. Darrieus to obtain a patent in France in 1926 for what is now known as the Darrieus wind turbine. We simply turn the horizontal rotation axis of the jumping rope by 90° to create a vertical rotation axis. Then, for good balancing, we connect two troposkein-shaped blades, and we have a vertical-axis wind turbine. In addition to the highly desirable pure tension force characteristic mentioned above, such a wind turbine has the additional attractive features of being able to operate regardless of the direction of the wind and having all of the necessary gear transmission machinery at ground level.

During the 1960s the National Research Council of Canada carried out extensive studies on Darrieus wind turbines, and in the late 1970s, the State Energy Commission of Western Australia initiated field investigation projects with these turbines for wind power generation. In 1973, Sandia National Laboratories in New Mexico began an extensive program of research and development on Darrieus wind turbines. A study by Blackwell and Reis (1974) gives a complete and concise analysis of the troposkein vertical-axis wind turbine.

Numerous field studies on Darrieus turbines of various sizes have been carried out by Sandia. Their large 34-meter vertical-axis turbine is seen in figure 14.4. Studies at this installation, located at Bushland, Texas, have been conducted in collaboration with the U.S. Department of Energy and Department of Agriculture.

The Bushland installation has a rotor diameter of 34 meters (i.e., $H = 17$ m) and a height of 50 meters (i.e., $L = 50$ m). When operating at the rated speed of 37.5 revolutions per minute (i.e., $\omega = 3.93$ radians per second), in a wind velocity of 12.5 meters per second (28 miles per hour), it is designed to produce 500 kilowatts (670 horsepower).

Looking upward, the turbine rotates clockwise. The blades are made of aluminum and have a cross section similar to an airfoil; the chord length is approximately 1.0 meters.

FIG. 14.4

The 34-meter vertical-axis wind turbine at Bushland, Texas. This Darrieus turbine, with $H = 17$ m and $L = 50$ m, is designed to produce 500 kW of power. (Photograph provided by Sandia National Laboratories.)

Darrieus Wind Turbine Problems

PROBLEM 1. The Sandia-DOE-USDA vertical-axis wind turbine at Bushland, Texas has the dimensions $H = 17$ m and $L = 50$ m. Show that the "swept area"—that is, the plane area within the boundary defined by the two troposkein blades—is $A_s = 1{,}105\ \text{m}^2$.

Hint. Use equation (14.17) to obtain the value $k = 0.450$. Then employ equation (14.15) to compute the blade profile. A numerical integration is needed to obtain A_s.

PROBLEM 2. Show that the length of each blade is $S = 62.4$ m.

PROBLEM 3. At the rated speed of 37.5 rpm ($\omega = 3.93$ rad / s), what is the magnitude of the blade tension, T_0? Assume that the mass of the blade is $\rho = 25$ kg / m.

Hint. From equation (14.14) you can establish that

$$T_0 = \frac{\rho\omega^2 H^2}{4}\left(\frac{1-k^2}{k^2}\right). \tag{14.19}$$

Answer. $S = 110{,}000$ newtons $= 11{,}200$ kg $= 24{,}700$ lb.

PROBLEM 4. The maximum blade tension, T_m, occurs at the junction of the blade and the axis. What is the value of T_m?

Answer. $T_m = 165{,}800$ newtons $= 16{,}900$ kg $= 37{,}300$ lb.

15

The Crisis of the Deficit: Gompertz to the Rescue

In preceding chapters we have seen how mathematics can be utilized to provide answers to many kinds of practical problems of the real world. Without question, among the most important categories of such problems are those associated with the economics and finance of our nation and our federal government.

In this regard, many people think we presently face the following very serious problem. The problem is that for quite a long time our government has been giving out money much faster than it has been taking in money. Nearly everyone believes that this is not a good idea.

Some say this is like a water storage tank. Water is being pumped into the tank but it is being drained out even faster; sooner or later the tank will be empty. Sure, the valve on the outlet could be partially or entirely closed; but what about all the people who need, or think they need, the full supply of water? Or more water could be pumped into the tank; but new sources of water are not easy to find.

What to do? Decrease the outflow? Increase the inflow? Very interesting phenomenon. Indeed, this so-called hydraulic analogy, much elaborated, was physically constructed years ago by Allen (1965), a noted British mathematical economist, to demonstrate how the Bank of England works.

With that as our prologue, we are now going to examine the federal debt, deficit, gross domestic product, and some other indices of economics. Then we will use a mathematical relationship devised by a fellow named Gompertz to try to bring things under control.

In chapter 2, we tabulated the amounts of the federal debt of the United States for the period from 1970 to 1992. For convenience, those data are shown here in table 15.1. In addition, the amounts of the annual gross domestic product (GDP) for the same period are listed. This information is plotted in figure 15.1.

We let N represent the size of the federal debt measured in billions of dollars. Looking at the N data in figure 15.1 (open circles), we make the assumption that the growth is exponential.

TABLE 15.1

The federal debt and the gross domestic product (GDP), 1970 to 1992

		Debt	*GDP*
Year	*t*	*(billion dollars)*	
1970	0	372.6	1,015.5
1972	2	427.8	1,215.5
1974	4	475.2	1,474.4
1976	6	621.6	1,784.8
1978	8	772.7	2,254.3
1980	10	908.7	2,732.0
1982	12	1,142.9	3,179.7
1984	14	1,573.0	3,801.5
1986	16	2,111.0	4,278.0
1988	18	2,586.9	4,908.4
1990	20	3,071.1	5,465.1
1992	22	4,077.5	5,834.2

Source: Data from U.S. Bureau of the Census (1994).

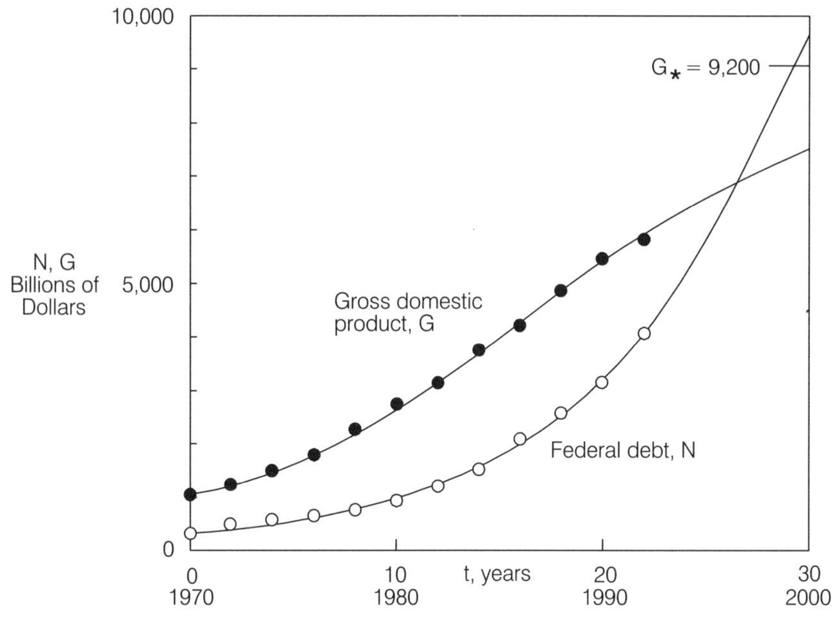

FIG. 15.1

Growth of the federal debt and the gross domestic product of the United States. The *N* curve is an exponential; the *G* curve is a logistic.

If so, we have

$$\frac{dN}{dt} = a_* N, \tag{15.1}$$

where N is the debt at time t and a_* is a growth coefficient. Integrating this expression gives the exponential growth equation

$$N = N_0 e^{a_* t}, \tag{15.2}$$

in which N_0 is the value of N at time $t = 0$ (i.e., the year 1970). If we take the logarithms of both sides of this equation, we obtain

$$\log_e N = \log_e N_0 + a_* t. \tag{15.3}$$

We know from analytic geometry that this is the equation of a straight line, $y = k_0 + k_1 x$. Using the numbers in the table, we carry out a least-squares computation to determine the numerical values: $N_0 = 326$ billion dollars and $a_* = 0.113$ per year. If these numbers are employed in equation (15.2), we produce the solid curve for the federal debt, N, in figure 15.1.

Similarly, we let G represent the gross domestic product in billions of dollars per year. The GDP data of table 15.1 are shown as the solid points in figure 15.1. This time, unlike the N data, it appears that G is *not* growing exponentially; the G data seem to be slowing down from about 1985 onward.

You will remember from chapter 3, that we introduced the equation that describes logistic growth:

$$\frac{dG}{dt} = aG\left(1 - \frac{G}{G_*}\right), \tag{15.4}$$

where G_* is the value of G as $t \to \infty$ and a is another growth coefficient. We shall try to fit a logistic curve to the G data.

The relationship of equation (15.4) is written in the form

$$\frac{1}{G}\frac{dG}{dt} = a - \frac{a}{G_*}G. \tag{15.5}$$

Again, in the vernacular of analytic geometry, this is the equation of a straight line. Writing this expression in so-called finite-difference form, utilizing the data of table 15.1, and performing another least-squares calculation gives the values $G_* = 9{,}200$ billion dollars per year and $a = 0.121$ per year.

A relatively easy problem in integral calculus involving equation (15.4) gives us the equation for logistic growth:

$$G = \frac{G_*}{1 + \left(\dfrac{G_*}{G_0} - 1\right)e^{-at}}, \tag{15.6}$$

in which G_0 is the value of G at $t = 0$ (i.e., 1970); from the data listed in the table and equation (15.6) we get $G_0 = 1{,}019$. Substituting these numbers into the equation gives the solid curve in figure 15.1 for the gross domestic product, G.

If you would like a bit of algebra, use equations (15.2) and (15.6) to show that the relationship between the federal debt, N, and the gross domestic product, G, is

$$N = N_0\left[m\left(\frac{G}{G_* - G}\right)\right]^{a_*/a}, \qquad (15.7)$$

where $m = (G_*/G_0) - 1$.

There is a very interesting intersection in figure 15.1: where the N and G curves cross, that is, where the federal debt is equal to the gross domestic product. You can easily determine that the intersection occurs at $t = 27.3$ (i.e., 1997) and $N = G = 7{,}150$. For the years beyond 1997, at the rate things are now going, the federal debt will increasingly exceed the gross domestic product. This should cheer you up.

The Gompertz Equation

In 1825 Benjamin Gompertz devised a mathematical expression that may help us get out of this unhappy situation. Now Gompertz, who lived in England, was a very clever actuary who devised tables and equations for use in the life insurance business. To be brief, he assumed that many things tend to grow exponentially but that the growth coefficient, a_*, is not always constant; it may change with time. Specifically, Gompertz assumed that the exponential growth relationship can be written in the form

$$\frac{dN}{dt} = a_*(t)N = a_{*0}e^{-ct}N, \qquad (15.8)$$

in which a_{*0} is a constant and c is a kind of decay coefficient. This expression indicates that when time $t = 0$, then $a_* = a_{*0}$. However, as t increases, the growth coefficient, a_*, gets smaller and smaller, and eventually it becomes zero.

Well, this is exactly what we need. With reference to equation (15.8), we want the time rate of change of the debt—that is, the deficit—to slow down and finally stop. A neat way to do this is to gradually reduce the growth coefficient a_*; this is precisely what the Gompertz relationship nicely provides.

Clearly, equation (15.8) offers a problem in integral calculus. You will be able to show that the equation for Gompertz growth is

$$N = N_0 e^{(a_{*0}/c)(1-e^{-ct})}, \tag{15.9}$$

where N_0 is the value of N when $t = 0$. (In a moment we are going to shift the time origin from 1970 to 1993. So be alert.)

The Federal Deficit

By definition, the deficit is equal to the time rate of change of debt: $D = dN/dt$. In our water storage tank, the excess of water outflow over inflow, measured in gallons per minute, corresponds to the excess of expenditures over revenues—that is, the *deficit* —in dollars per year. The consequent reduction of water storage in the tank, measured in gallons, corresponds to the reduction of assets—or increase of *debt*—measured in dollars.

So here is a little problem in differential calculus. If you differentiate the expression for the debt, equation (15.9), you obtain the following equation for the deficit, in which $D_0 = a_* N_0$:

$$D = D_0 e^{(a_{*0}/c)(1-e^{-ct})-ct}. \tag{15.10}$$

This is an interesting and useful result. You will note that as t increases then D decreases and as $t \to \infty$ then $D \to 0$. This is the kind of relationship that will reduce and eventually eliminate the annual budget deficit we now have.

Wherein We Abolish the Deficit

The federal government announced that commencing in 1993 it began to reduce the annual deficit at a rate sufficient to attain one-half the 1993 amount by 1998 with continued deficit reduction to zero in ensuing years. This is good news. Now we can really start to solve the problem.

The equation for computing the present federal debt N is given by equation (15.2). So the equation for the current deficit $D = dN/dt$ is easily determined. Thus we have

$$N = N_0 e^{a_{*0} t}; \qquad D = D_0 e^{a_{*0} t}; \qquad D_0 = a_{*0} N_0. \qquad (15.11)$$

Some comments. The quantity a_{*0} indicates that the growth coefficient is constant up to 1993. Until then, the time origin, $t = 0$, corresponds to 1970. Substituting the numerical values $N_0 = 326$ billion dollars, $a_{*0} = 0.113$ per year, and $t = 23$ yr (i.e., 1993–70) in equation (15.11) gives, for 1993, $N = 4{,}385$ billion and $D = 495$ billion/yr. Now we shift the time origin, $t = 0$, to 1993. Consequently, the revised "initial conditions" are $N_0 =$ 4,385 and $D_0 = 495$.

As yet we do not know the numerical value of the decay coefficient, c. However, the government tells us it wants to cut the deficit in half after five years. Accordingly, we substitute $D = 495/2 = 247.5$ and $t = 5$ into equation (15.10), which, of course, is the Gompertz model for deficit reduction. Iterated solutions of the resulting equation easily provides $c = 0.209$/yr.

Finally, with these numerical values, we use equation (15.9) and (15.10) to compute the debt, N, and deficit, D, for the period commencing in 1993. The resulting curves are shown in figure 15.2; calculated amounts are shown in table 15.2. For comparison, amounts are also listed that are based on exponential growth from 1970 onward.

We conclude that our Gompertz reduction scheme is fairly successful. The annual deficit will be reduced to 248 billion dollars per year by 1998, to 174 billion by 2000, and to 66 billion by 2005. This is still a lot of money but look what it would have been with continued exponential growth.

Of course, the debt will continue to increase. It will reach 6,541 billion dollars by 2000 and 7,201 billion by 2005. Ultimately, it will level off at 7,530 billion. However, with continuation of exponential growth, the federal debt would have reached absolutely astronomical magnitudes following the turn of the century.

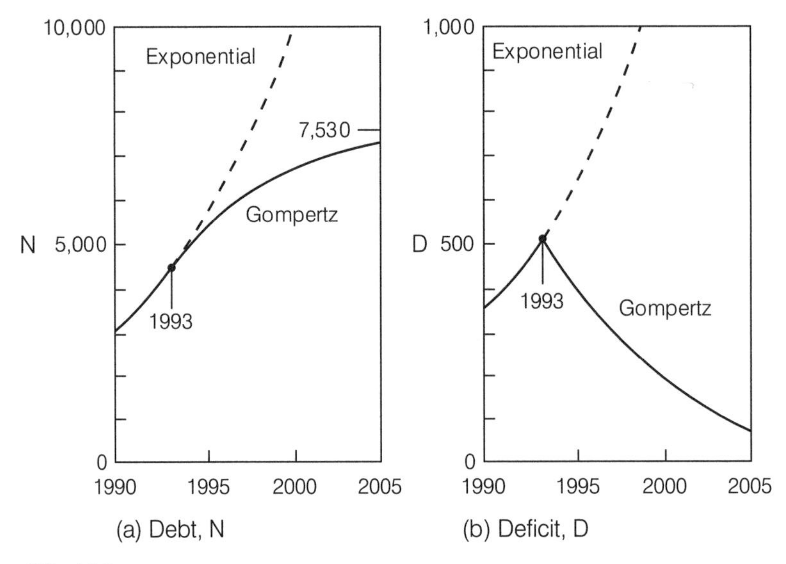

FIG. 15.2

Debt and deficit curves, computed from the Gompertz equation, for the period commencing in 1993

TABLE 15.2

Amounts of the debt and deficit, computed from the Gompertz equation and from the exponential equation, for years commencing in 1993

		Federal debt, N billion dollars		*Federal deficit, D billion dollars/year*	
Year	*t*	*Gompertz*	*Exponential*	*Gompertz*	*Exponential*
1993	0	4,385	4,385	495	495
1995	2	5,274	5,497	392	621
1998	5	6,223	7,715	248	872
2000	7	6,541	9,671	174	1,013
2005	12	7,201	17,016	66	1,923
2010	17	7,409	29,938	24	3,383

So the *Gompertz distribution* has provided a simple and logical way out of this financial mess. This distribution is very useful, not only in economics but also in many areas of science and technology. For example, industrial engineers have used the Gompertz distribution in connection with technology transfer and technology substitution. Geographers and sociologists have employed it in phenomena of innovation diffusion. Medical researchers have utilized it as a framework to characterize certain types of cancer growth. A number of applications of the Gompertz distribution are presented by Banks (1994).

PROBLEM. It is not hard to show that for logistic growth, the ordinate of the inflection point of the N curve is exactly one-half of the ordinate of the equilibrium value or carrying capacity. That is, $N_i / N_* = 1/2$. If you would like to analyze an interesting problem in differential calculus, you will be able to prove that for Gompertz growth, $N_i / N_* = 1/e$.

More Information about the Gross Domestic Product

By definition, the gross domestic product (GDP) is the total value of all the goods and services produced within a nation during a one-year period. Based on the GDP data shown in table 15.1 and figure 15.1, we obtained a logistic-type growth of America's GDP over the period from 1970 to 1992. Our equation was

$$G = \frac{G_*}{1 + me^{-at}}, \tag{15.12}$$

in which $m = (G_* / G_0) - 1$, $G_0 = 1{,}019$ billion dollars/yr, $G_* = 9{,}200$ billion dollars/year, and $a = 0.121$ yr.

So here is a good question. What is the value of all the goods and services produced by the United States over a specific period of years? In other words, what is the magnitude of the *cumulative gross domestic product* (CGDP) from, say, 1970 to 1990, or from 1970 to 2000?

This presents an interesting problem in integral calculus; we simply need to integrate equation (15.12). We let W represent

TABLE 15.3

Computed and measured values of the cumulative gross domestic product (CGDP) in billion dollars

1970 to year	*CGDP Computed*	*CGDP Measured*
1975	7,537	7,772
1980	18,915	19,056
1985	36,155	36,589
1990	60,936	61,030
1995	90,804	
2000	126,881	

the CGDP and write

$$W = \int_0^t G\,dt = \int_0^t \frac{G_*}{1 + me^{-at}}\,dt. \tag{15.13}$$

Integration of this equation gives

$$W = \frac{G_*}{a}\left[at + \log_e\left(\frac{1 + me^{-at}}{1 + m}\right)\right]. \tag{15.14}$$

Incidentally, in order to make a direct comparison of the computed values of W and the summarized measured annual data, it is necessary to set the time origin, $t = 0$, at 1969.5. Finally, substitution of numerical values into equation (15.14) yields the CGDP quantities shown in table 15.3.

We conclude our examination of this topic with some startling numbers. From table 15.3 it is noted that the cumulative gross domestic product (CGDP) of the United States for the twenty-year period 1970 through 1990 was approximately *61 trillion dollars*. Furthermore, our mathematical model predicts that the CGDP for the thirty-year period 1970 to 2000 will be around *127 trillion*.

16

How to Reduce the Population with Differential Equations

It seems as if everything grows: The trees in the park, your monthly fuel bill, people, your taxes, the federal debt, the amount of pollutants in the atmosphere, the number of cars on the freeway, the populations of cities, nations, and the world, and so on. Sometimes things grow at a steady rate and sometimes they speed up or slow down. Fuel bills rise fast with the onset of winter and decline in the spring; the reverse is true of plants and trees. Everyone knows that teenagers grow at incredibly high speeds, thirty-somethingers have leveled off, at least in height, and old timers, to coin a phrase, are going downhill. Suburban towns grow rapidly; central cities shrink slowly.

Using Mathematics to Describe Growth

It is possible to describe these various growth phenomena with mathematical models; some are simple and some are complicated. The most famous example is the familiar Malthusian or exponential growth model. In differential equation form, it has the definition

$$\frac{dN}{dt} = aN, \tag{16.1}$$

in which N is the magnitude of a growing quantity, t is time, and a is a growth coefficient. The solution to this equation is

$$N = N_0 e^{at}, \tag{16.2}$$

where N_0 is the value of N when time $t = 0$.

Virtually all mathematical models of growth can be defined as variations of equation (16.1). To put these models into a standard form, we rewrite this equation as

$$\frac{1}{N}\frac{dN}{dt} = a(t). \tag{16.3}$$

The quantity on the left is called the specific growth rate. It is equal to the growth coefficient $a(t)$, which usually depends on time, t. Here are some examples of growth models:

Constant Growth Coefficient

$$a(t) = a_0 = \text{constant}. \tag{16.4}$$

This, of course, is exponential growth; the solution is equation (16.2).

Linearly Decreasing Growth Coefficient

$$a(t) = a_0(1 - mt), \tag{16.5}$$

in which m is a positive constant. This example is interesting and useful; it leads to a *normal* or *Gaussian* distribution of N.

Sinusoidally Variable Growth Coefficient

$$a(t) = a_0 + a' \sin \omega t. \tag{16.6}$$

This equation indicates that the growth coefficient oscillates sinusoidally with frequency, ω. The average value of the growth coefficient during an oscillation is a_0; the amplitude of the oscillation is a'. This model could be used to describe the growth of a tree; it speeds up during the spring and summer and slows down during autumn and winter. In this case, $\omega = 2\pi/T$, where the period of oscillation $T = 365$ days.

Exponentially Decreasing Growth Coefficient

$$a(t) = a_0 e^{-kt}. \tag{16.7}$$

This equation defines so-called Gompertz growth, which we used in chapter 15.

In these examples, it is indicated that the growth coefficient depends on the *independent* variable, t; that is, $a = a(t)$. Sometimes, for one reason or another, the growth coefficient is defined in terms of the *dependent* variable, N; that is, $a = a(N)$. In the long run, it doesn't matter which way we do it. Some examples in this category are the following:

Logistic Growth Coefficient

$$a(N) = a_0 - bN = a_0\left(1 - \frac{N}{N_*}\right), \tag{16.8}$$

where b is a "crowding coefficient" and N_* is the equilibrium value or carrying capacity. Clearly, $N_* = a_0/b$.

Coalition Growth Coefficient

$$a(N) = a_0 N. \tag{16.9}$$

Modified Coalition Growth Coefficient

$$a(N) = a_0 N\left(1 - \frac{N}{N_*}\right). \tag{16.10}$$

There are many other expressions for growth coefficients, in the form $a(t)$ or $a(N)$, but these are sufficient for now.

Effects of Poison or Pollution on Growth

Years ago there was much interest in using the logistic equation as a mathematical framework for analysis of all kinds of growth processes, ranging from the growth of sunflower plants to that of the population of the United States. For example, biologists and bacteriologists performed laboratory experiments to try to demonstrate that microorganisms grow logistically. We know that with logistic growth, the concentration of bacteria in a test tube increases rapidly—indeed, exponentially—at the start of an experiment, then gradually slows down and finally levels off at a constant value, called the equilibrium concentration or carrying capacity. Well, all too frequently in these early experiments, the bacteria concentration, after a rapid initial rise, did not level off at a constant value but instead reached a peak concentration and then plunged to zero. This "rise and decline" happened even though there were sufficient nutrients in the test tube to maintain an equilibrium concentration of the bacteria. Researchers were unable to explain the strange behavior.

To make a long story short, two famous mathematicians became interested in this problem and analyzed what was going on. The first of these was Vito Volterra (1860–1940), perhaps the greatest mathematician ever produced by Italy. He is most famous for his work on integral equations and the calculus of variations. The second was Vladimir Kostitzin (1882–1963), born in Russia but a resident of France for most of his life. Though highly competent in many fields of science, his most noteworthy contributions were in theoretical ecology.

During the 1930s, when both were living in Paris, Volterra and Kostitzin produced a series of publications—nicely provided for us by Scudo and Ziegler (1978)—on many subjects of mathematical biology. Among these were several dealing with the above problem of bacterial "crashing." They explained that the decline to extinction was due to contamination of the growth environment by lethal products generated by the bacteria themselves, or, in other words, self-poisoning of the system.

The equation that Volterra and Kostitzin devised and used to describe this growth process is the following:

$$\frac{1}{N}\frac{dN}{dt} = a_0 - bN - c\int_0^t N\,dt. \tag{16.11}$$

Let us examine this equation term by term. The left-hand side is the specific growth rate as we arranged it earlier. The first term on the right-hand side is the growth coefficient, a_0. If the following two terms were absent (i.e., if $b = 0$ and $c = 0$), we would obviously have exponential growth.

The second term on the right describes the effect of crowding. It says that the crowding effect is directly proportional to the magnitude of N and, because of the minus sign, reduces the size of the growth coefficient a_0. If the following, final term were absent (i.e., if $c = 0$), we could have logistic growth.

The third term on the right represents the effect of self-poisoning. The appearance of the integral indicates that this self-poisoning began at time $t = 0$, and has *accumulated* from then to the present time t. It is clear that this term also reduces the magnitude of the growth coefficient a_0.

To paraphrase the observations of Kostitzin himself, it can be said that equation (16.11) represents the growth of a population that is intoxicated by its own catabolic products. In other words, it mathematically describes a process leading to self-destruction.

It happens that equation (16.11) is an example of an *integro-differential equation.* It is called this because the dependent variable, N, appears in a *derivative* term and also in an *integral* term.

This equation is fairly easy to solve. However, in general, integro-differential equations are very difficult to handle. No mathematician, past or present, has contributed more than Volterra to our knowledge of how to deal with such equations. They are extremely important; they appear in a great many problems in applied mathematics, engineering, and science.

The Solution to Our Integro-Differential Equation

To solve our equation easily, we make a substantial simplification. From here on we assume that $b = 0$ in equation (16.11). This means that there is no crowding effect; the carrying capacity, $N_* = a/b$, is infinitely large. In other words, there is no limitation of resources. Accordingly (dropping the subscript on the growth coefficient), equation (16.11) becomes

$$\frac{dN}{dt} = aN + cN\int_0^t N\,dt. \tag{16.12}$$

Next, we make the substitution

$$P = \int_0^t N\,dt \qquad \text{and so} \qquad \frac{dP}{dt} = N. \tag{16.13}$$

Putting the first of these expressions into equation (16.12) gives

$$\frac{dN}{dt} = aN - cNP. \tag{16.14}$$

So now we have replaced our integro-differential equation, equation (16.11), with the following two differential equations:

$$\frac{dN}{dt} = aN - cNP \qquad \text{and} \qquad \frac{dP}{dt} = N. \tag{16.15}$$

Dividing the first of these by the second gives

$$\frac{dN}{dP} = a - cP. \tag{16.16}$$

In a form ready for integration this equation becomes

$$\int_{N_0}^{N} dN = \int_0^P (a - cP)\,dP. \tag{16.17}$$

The lower limits on the integrals state that at time $t = 0$, the amount of poison—or pollution, if you prefer—is $P = 0$ and the number of bacteria—or people, if you like—is $N = N_0$.

Accordingly, integrating this equation gives

$$N = N_0 + aP - \tfrac{1}{2}cP^2. \qquad (16.18)$$

This expression relates the two dependent variables, N and P; time, t, of course, is the independent variable.

PROBLEM 1. Research bacteriologists conducted experiments with *Bacillus coli* for the purpose of measuring growth rates, maximum concentrations, and extinction times. The experiments were carried out in saline solutions, held at 37°C, with various initial amounts of peptone nutrient.

In one experiment they determined that $a = 1.15$, $c = 2.0 \times 10^{-8}$, and $N_0 = 0.5 \times 10^6$ bacteria / cm^3. The experiment ran for about ten hours.

How much poison was emitted by the bacteria and accumulated in the test tube during the entire experiment? That is, what was the value of P when the bacterial concentration, N, after reaching a maximum, fell to zero?

Hint. To answer the question you must compute P from equation (16.18). This is a quadratic equation. Show that

$$P = \frac{a}{c}\left[1 + \sqrt{1 - \frac{2c(N - N_0)}{a^2}}\right]. \qquad (16.19)$$

Then set $N = 0$, substitute the above numbers, and obtain the result: $P = 115$ million bacteria-hours of poison emission and accumulation.

The Final Equations for the Population and Pollution

The result of equation (16.18), relating the magnitude of the growing quantity, N, to the size of the accumulated poison, P, is interesting. However, our goal is to determine N and P as functions of time, t. So using equation (16.18) and the second of equation (16.15), that is $N = dP/dt$, we obtain

$$\int_0^t dt = \int_0^P \frac{1}{N_0 + aP - \frac{1}{2}cP^2}\, dP. \qquad (16.20)$$

To evaluate the right-hand member of this equation we need a table of integrals. Then, with a bit of algebra, we obtain the answer

$$P = \frac{a}{c}\left[1 + \frac{a_*}{a}\tanh\left(\frac{1}{2}a_* t - \phi\right)\right], \tag{16.21}$$

in which

$$a_* = \sqrt{a^2 + 2cN_0} \qquad \text{and} \qquad \phi = \operatorname{arctanh}\frac{a}{a_*}. \tag{16.22}$$

Finally, again using $N = dP/dt$, differentiation of equation (16.21) yields

$$N = \frac{a_*^2}{2c}\operatorname{sech}^2\left(\frac{1}{2}a_* t - \phi\right). \tag{16.23}$$

PROBLEM 2. Show that the *maximum point* of the N curve has the coordinates

$$t_m = \frac{2\phi}{a_*}; \qquad N_m = \frac{a_*^2}{2c}. \tag{16.24}$$

Show also that the *inflection point* of the N curve and the slope at that point are given by

$$t_i = \frac{2}{a_*}\left(\phi \pm \operatorname{arctanh}\frac{1}{\sqrt{3}}\right); \qquad N_i = \frac{2}{3}N_m;$$
$$\left(\frac{dN}{dt}\right)_i = \pm\frac{2}{3\sqrt{3}}a_* N_m. \tag{16.25}$$

PROBLEM 3. For the *Bacillus coli* experiment of problem 1, plot the curves $N(t)$ and $P(t)$. In addition, compute the coordinates of (a) the maximum point and (b) the inflection point of the N curve.

Answers. (a) $t_m = 4.82$ hours, $N_m = 33.6 \times 10^6$ bacteria; (b) $t_i =$ 3.69 and 5.96 hrs, $N_i = 22.4 \times 10^6$ bacteria, $(dN/dt)_i = 15.0 \times 10^6$ bacteria / hour.

Finally, for the inflection point of the P curve, confirm the following;

$$t_i = \frac{2\phi}{a_*}; \qquad P_i = \frac{a}{c}; \qquad \left(\frac{dP}{dt}\right)_i = \frac{a_*^2}{2c}. \tag{16.26}$$

In Which We Reduce, Even Eliminate, the World's Population

Many believe that there are already too many people in the world. Maybe so, maybe not. In any event, during the coming decades, the world's population will certainly continue to increase. Then it will begin to level off. Following this, and because of accumulated global pollution, the population will start to decrease. We shall use our mathematical model, which resulted in equations (16.21) and (16.23), to examine the effects of this accumulated pollution on the size of the population.

This is not exactly a new point of view. It was the main thrust of studies carried out years ago at the Massachusetts Institute of Technology on "world dynamics" and "limits to growth"; the results of these investigations are presented by Forrester (1971) and Meadows (1974). Along the same lines is a publication entitled "Limits to Growth from Volterra Theory of Population" by Borsellino and Torre (1974); it is quite relevant to our problem.

By "pollution" we mean far more than simply air and water pollution and solid waste accumulation. We include deforestation, desertification, soil erosion, resource depletion, urban and rural decay, congestion, global warming, ozone destruction, polar ice cap melting, sea level rising, destruction of animal and plant species, nuclear damage, electromagnetic radiation, food poisoning, noise pollution, disease and epidemics—and you take it from there. Forrester put many of these factors into something he called a "quality of life index." Essentially, this is what we have in mind when we refer to our self-poisoning or self-polluting variable or parameter, P.

What we are now going to do, in an analytical way, is to show that the future population of the world, N, will be strongly

TABLE 16.1

Population of the world, 1950 to 1990, in billions of persons

Year	*t*	*N*
1950	0	2.565
1960	10	3.050
1970	20	3.721
1980	30	4.476
1990	40	5.320

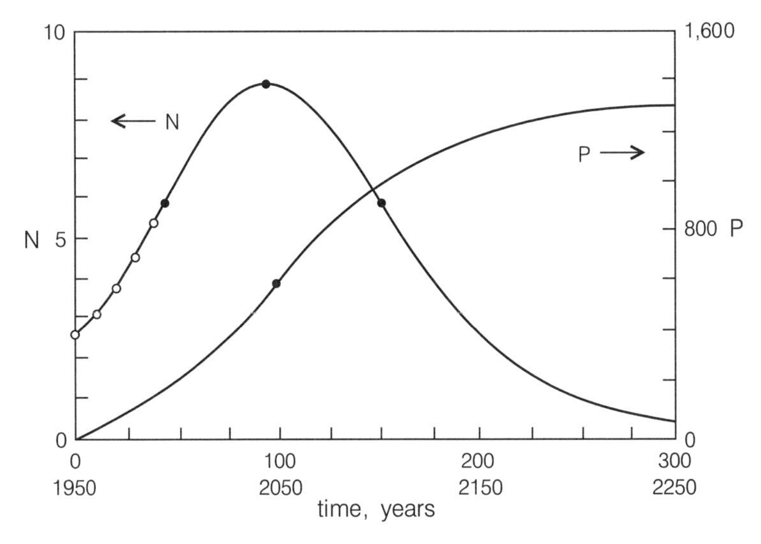

FIG. 16.1

Plots of world population, *N*, in billions of persons and accumulated pollution, *P*, in billions of person-years

affected—indeed, controlled—by the accumulated pollution of the world, *P*. We shall use the mathematical growth model developed above to accomplish this.

To begin with, we need the population data shown in table 16.1. The five points listed in the table are plotted in figure 16.1. Time $t = 0$ corresponds to 1950. For the moment, we disregard the solid curves in the figure.

In the equations we note that there are three constants that need to be determined: a, c, and N_0. From the data we have for years 1950 through 1990, we evaluate two of these, a and N_0. However, there is no way to determine the value of the pollution coefficient, c. Therefore, in order to get answers to our problem, it is necessary to make one assumption about the world's population. Specifically, we shall assume that the maximum population will occur in 2050, that is, at $t = 100$.

With this assumption, and using equation (16.24) for the time to maximum N, that is, $t_m = 2\phi/a_*$, we determine a_*. From equation (16.22) we calculate the values of c and ϕ. Finally, from equation (16.24) we compute the maximum population, N_m. The results are as follows: $a = 0.0206$ 1/year, $N_0 = 2.56 \times 10^9$ persons, $a_* = 0.0245$ 1/year, $c = 3.41 \times 10^{-14}$ 1/person-year^2, $\phi = 1.227$, and $N_m = 8.80 \times 10^9$ persons. We now have all the information needed to plot the N and P curves. From equations (16.21) and (16.23), the solid curves shown in figure 16.1 are constructed.

The inflection points of the population curve are given by equation (16.25). The values determined are $t_i = 46.3$ and 153.7 years, corresponding to 1996 and 2104, $N_i = 5.87 \times 10^6$ persons, and $(dN/dt)_i = 8.30 \times 10^6$ persons/year. These points are shown as the small solid dots on the N curve. Our growth model indicates that in 1996 (2104) the world's population will be increasing (decreasing) at a maximum rate of 83 million persons per year. Most importantly, the model also forecasts a maximum population of 8.80 billion persons in 2050. After that, as seen in the figure, the population will begin to decline.

The total amount of accumulated pollution between 1950 and 2000 will be about 212 billion person-years. For the time being, we shall have to settle for "person-years" as the suitable unit for pollution. If we had time, we could probably correlate this unit to Forrester's quality-of-life index, to equivalent tons of garbage, to a virtual economic cost, or whatever.

In any event, the total amount of pollution accumulated between 1950 and 2050, the year of maximum population, will be 604 billion person-years. Between 1950 and, say 2500—and forever thereafter since there are no people—it will be 1,322 billion person-years. Table 16.2 lists values of N and P for the years ahead.

TABLE 16.2

Projections of population, N, in billions of persons and accumulated pollution, P, in billions of person-years

Year	t	N	P
1950	0	2.6	0
1990	40	5.3	154
2000	50	6.2	212
2025	75	8.0	391
2050	100	8.8	604
2100	150	6.2	996
2150	200	2.6	1,208
2200	250	0.8	1,287

17

Shot Puts, Basketballs, and Water Fountains

One of the most ancient problems involving mathematical analysis is the problem of determining the trajectories—the flight paths—of projectiles. Historically, this branch of science has been known as *ballistics*. It is impossible to determine when mankind first became interested in the subject. It is safe to say that prehistoric cavemen, shooting rocks from slingshots and throwing spears at saber-toothed tigers, did not worry too much about the physics and mathematics involved. However, without doubt, the military engineers of Rome had extensive empirical knowledge concerning how high and how far their catapults could throw big stones. Likewise, the artillery specialists of the Middle Ages and later must have had fairly accurate information about the distance a particular mortar could heave a particular cannon ball.

Ballistics began to become a science after Galileo, Kepler, and, most importantly, Newton made their monumental contributions in the sixteenth and seventeenth centuries. During the many years to follow, and indeed up to the present time, mankind's eternal need to improve its weapons of war has resulted in incredible advances in the mathematical analysis of projectile trajectory. By the end of the first half of the twentieth century, the ballistics of projectiles—artillery shells and bombs—had been pretty well worked out. During the second half of the current century, we have seen that trajectory analysis for flight in space has met with phenomenal success.

Space flight trajectory analysis is an extremely complex problem. It must take into account the rotation of the earth, changes in gravitational forces and atmospheric conditions, changes in the mass of the projectile and its propelling force, and many other things. If you are interested in space trajectory analysis, the books by Bate et al. (1971) and Thompson (1986) are good places to start.

Well, we are not going to get that complicated. Furthermore, we are not interested in catapulting stones, shooting cannon balls, or dropping bombs. We are, as they say, sports-minded. So we shall analyze the trajectories of things like basketballs, shot-put shots, baseballs, and golf balls. In this chapter and those that follow we examine the trajectory problem when only gravity force is considered and also when both gravity force and air resistance are taken into account.

In addition, we shall analyze the closely related phenomenon of water jet trajectories. It turns out that, in the absence of air resistance, the trajectory of a basketball, for example, and the graceful curve of a water jet are the same; both are parabolas.

When Can We Neglect Air Resistance?

Even the relatively simple problem of trajectory analysis of "sporting balls" is not an easy one. The reason: air, through which the ball must travel, exerts a drag force—and sometimes a lift force—on the ball; this greatly complicates matters. The drag force, or air resistance, is the one we now consider. If we could somehow ignore the effect of air resistance our problem would be considerably simplified. Let us look into this.

Suppose that we toss a tennis ball vertically upward with an initial velocity U_0. Using magic, we remove all the air around us, and so, in this vacuum, the only force acting on our tennis ball is its weight, that is, the force of gravity.

Utilizing Newton's law of motion, we obtain the relationship

$$-W = m\frac{dU}{dt} = \frac{W}{g}U\frac{dU}{dy}, \tag{17.1}$$

where W is the weight of the object (we insert the minus sign because the weight acts in the direction opposite to the direction of motion), $m = W/g$ by definition, g is the force due to gravity ($g = 32.2\ \text{ft/s}^2 = 9.82\ \text{m/s}^2$), and U is the velocity of the object at any height y above the point of release, $y = 0$. The release or initial velocity is U_0.

We easily solve equation (17.1) to obtain

$$U = \sqrt{U_0^2 - 2gy}\,. \tag{17.2}$$

Clearly, when the object reaches its maximum height, y_m, the velocity at that instant is $U = 0$. So from equation (17.2) we get

$$y_m = \frac{U_0^2}{2g}. \tag{17.3}$$

For example, if we throw our tennis ball vertically upward in a vacuum with an initial velocity $U_0 = 100$ ft/s, it will reach a maximum height $y_m = 155$ ft.

The next problem is a lot more complicated. Instead of a vacuum, suppose that now our ball moves through air. So now we have two external forces acting on the ball: weight and air resistance. In this case, equation (17.1) becomes

$$-W - \frac{1}{2}\rho_a C_D A U^2 = \frac{W}{g}U\frac{dU}{dy}, \tag{17.4}$$

in which the second term in the left-hand member is the force due to air resistance; we call this the *drag*. The quantity ρ_a is the density of air, C_D is a so-called drag coefficient, and A is the projected area of the object; for a sphere, $A = (\pi/4)D^2$, where D is the diameter.

Homework Problems

Here are a few introductory problems dealing with the vertical trajectory of a sphere (e.g., a tennis ball) in air.

PROBLEM 1. This first problem involves integral calculus. Show that the solution to equation (17.4) yields

$$y_m = \frac{U_0^2}{2g}\frac{1}{n}\log_e(1+n), \tag{17.5}$$

where y_m is the maximum height a sphere will reach in air with an initial velocity U_0, and

$$n = \frac{3\rho_a C_D U_0^2}{4\rho g D}, \tag{17.6}$$

in which ρ_a is the density of air and ρ is the density of the sphere.

PROBLEM 2. Show that this quantity, n, is the ratio of the initial drag on the sphere to its weight.

PROBLEM 3. In equation (17.6), if ρ_a or C_D is zero (so that the drag is zero and hence $n = 0$), show that equation (17.5) reduces to equation (17.3).

PROBLEM 4. A new tennis ball has a diameter $D = 2.52$ in $= 0.210$ ft, and weight $W = 0.128$ lb. Accordingly, its average specific weight is $\gamma = W/(\pi/6)D^3 = 26.4$ lb/ft^3, and so its density is $\rho = \gamma/g = 0.820$ slugs/ft^3. An average value of the drag coefficient is $C_D = 0.30$.

The density of air (at sea level pressure and 70°F temperature) is $\rho_a = 0.0024$ slugs/ft^3. If the initial velocity of the tennis ball is $U_0 = 100$ ft/s, show that $n = 0.97$. Using equation (17.5), show that the ball will reach a maximum height $y_m = 108$ ft. This is substantially less than the earlier result, $y_m = 155$ ft, in which drag was neglected.

We have seen that the quantity n describes the relative influence of air resistance. If n is "small" then we can neglect air resistance and if n is "large" we cannot.

TABLE 17.1

Numerical values of the drag-weight ratio n for several sporting balls

Quantity	*Shot-put shot*	*Basketball*	*Baseball*	*Tennis ball*	*Golf ball*
Diameter D, ft	0.361	0.796	0.241	0.210	0.138
Weight W, lb	16.0	1.323	0.320	0.128	0.101
Density ρ, slug/ft^3	20.20	0.156	1.356	0.820	2.279
Velocity U_0, ft/s	45	25	140	150	230
$n = \dfrac{3\rho_a C_D U_0^2}{4\rho g D}$	0.005	0.085	1.01	2.19	2.83

Note: $\rho_a = 0.0024$ slug/ft^3; $C_D = 0.30$; $g = 32.2$ ft/s^2.

Numerical values of the diameters, weights, typical maximum velocities, and drag-weight ratios, n, for several sporting balls are listed in table 17.1. It can be seen in the table that the magnitude of n ranges from $n = 0.005$ for a shot-put shot and $n = 0.085$ for a basketball to $n = 2.83$ for a golf ball. On this basis, it is reasonable to neglect air resistance in the trajectory analysis of shot-put shots and basketballs (at least for free throws for which the velocity, U_0, is relatively small). However, because of the large values of n, we cannot ignore air resistance when computing the trajectories of golf balls, tennis balls, and baseballs.

Projectile Trajectories: Launch Point and Impact Point Are at the Same Elevation

From here on, in this chapter, we are going to neglect air resistance. It will be assumed that our projectile—be it a ball, a paper wad, an arrow, or a water jet—is moving through a vacuum. This assumption makes trajectory analysis pretty easy. In ensuing chapters, in which we consider the flight paths of baseballs and golf balls, we shall examine the much more difficult problem in which the projectile is moving through the air.

We start with the problem in which the launch point and the impact point are at the same elevation. The definition sketch of this problem is seen in figure 17.1. The projectile is launched (i.e., tossed, thrown, fired) from the origin at an angle α to the horizontal with a velocity U_0.

In the absence of air resistance, the coordinates of any point, $P(x, y)$, on the projectile's trajectory are

$$x = (U_0 \cos \alpha)t; \qquad y = (U_0 \sin \alpha)t - \tfrac{1}{2}gt^2, \tag{17.7}$$

where t is the time elapsed since launch. The values of x (abscissa) and y (ordinate) are expressed in so-called parametric form in which t is the parameter.

We solve the first of equations (17.7) to get $t = x/U_0 \cos \alpha$. Then we substitute this into the second equation, to obtain

$$y = (\tan \alpha)x - \frac{g}{2U_0^2}(\sec^2 \alpha)x^2. \tag{17.8}$$

This is the equation of the projectile's trajectory. You will observe that equation (17.8) is a parabola.

Now for some questions. First, for given values of U_0 and α, how far will the projectile go? That is, what is the range, L? From figure 17,1., when $x = L$ then $y = 0$. Substituting these values into equation (17.8) gives the answer

$$L = \frac{2U_0^2}{g} \sin \alpha \cos \alpha. \tag{17.9}$$

Next question: What is the maximum height of the projectile trajectory? Clearly, this is simply a problem in differential calculus. Differentiating equation (17.8) with respect to x and setting the result equal to zero gives the expression

$$x_m = \frac{U_0^2}{g} \sin \alpha \cos \alpha. \tag{17.10}$$

where, as seen in figure 17.1, x_m is the distance along the x-axis corresponding to maximum height. Finally, substituting equation

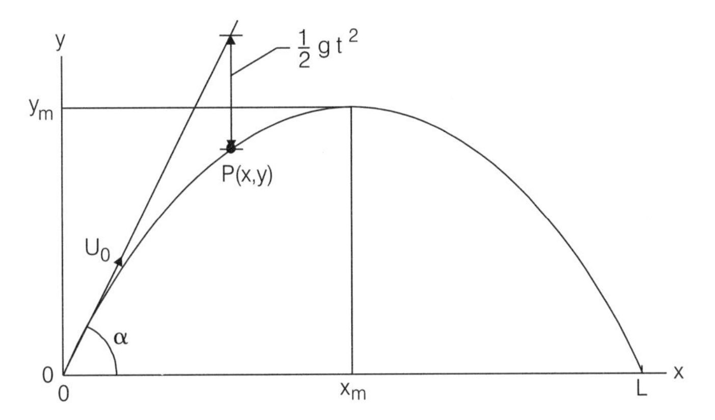

FIG. 17.1

Definition sketch for projectile trajectory analysis. Launch point and impact point are at the same elevation.

(17.10) into equation (17.8) gives

$$y_m = \frac{U_0^2}{sg} \sin^2 \alpha. \tag{17.11}$$

This is the maximum height of the trajectory.

Another question: For a specified initial velocity U_0, is there a launch angle α that provides a maximum range? The solution to this problem is obtained by differentiating equation (17.9) with respect to α and equating the result to zero. The answer is $\alpha_m = 45°$. Thus, maximum range is obtained when the launch angle is 45°. Substituting this into equations (17.9) and (17.11) gives

$$L_m = \frac{U_0^2}{g} \quad \text{and} \quad y_m = \frac{U_0^2}{4g} \quad (\alpha = 45°). \tag{17.12}$$

Finally, how long does it take for the projectile to go from $x = 0$ to $x = L$, that is, what is the flight time (or, if you prefer, the "hang time"), T? The answer to this question is provided by the first of equations (17.7). We easily obtain $T = L/U_0 \cos \alpha$. If

$\alpha = 45°$, the flight time is

$$T = \sqrt{2}\, L_m / U_0 \qquad (\alpha = 45°). \tag{17.13}$$

Let us use some numbers. Suppose that a ball is thrown with an initial velocity $U_0 = 75$ ft/s at an angle $\alpha = 45°$. From equations (17.12), the horizontal length of flight of the ball (i.e., the range) is $L_m = 174.7$ ft, and the maximum height is $y_m = 43.7$ ft. The flight time, computed from equation (17.13), is $T = 3.3$ s.

Projectile Trajectories: Launch Point and Impact Point Are at Different Elevations

Next, we examine the slightly more difficult problem in which the launch point and impact point are at different elevations. As shown in figure 17.2, the impact point may be a distance H *above* the launch point (e.g., when a basketball player is shooting a free throw). Alternatively, also indicated in the figure, the impact point may be a distance H *below* the launch point (e.g., when a shot-putter is putting a shot).

Analysis of this problem gives the same equation as before for the projectile trajectory. That is,

$$y = (\tan \alpha) x - \frac{g}{2U_0^2} (\sec^2 \alpha) x^2. \tag{17.14}$$

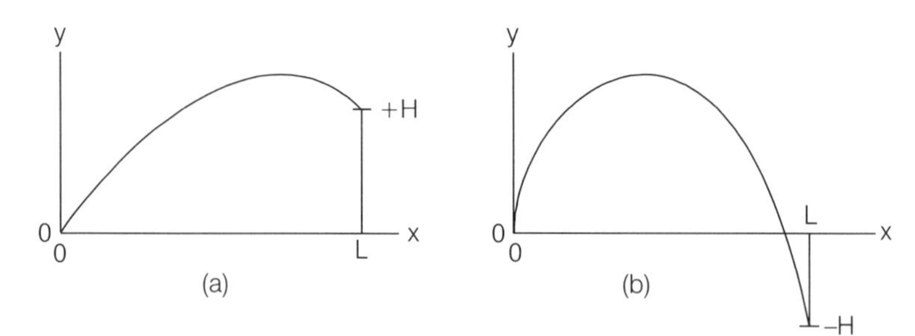

FIG. 17.2

Definition sketch for projectile trajectory analysis. Impact point (*a*) above and (*b*) below launch point.

From the definition sketches, it is clear that when $x = L$, then $y = +H$ and $y = -H$, respectively. Substituting these values into equation (17.14) gives a quadratic equation in L, whose solution is

$$L = \frac{U_0^2}{g} \sin\alpha\cos\alpha \left[1 \pm \sqrt{1 - \frac{2gH}{U_0^2 \sin^2\alpha}} \right], \tag{17.15}$$

where H may be positive (shooting uphill) or negative (shooting downhill). As we would expect, if $H = 0$, this reduces to equation (17.9).

Again, the coordinates of the maximum height point are

$$x_m = \frac{U_0^2}{g} \sin\alpha\cos\alpha; \qquad y_m = \frac{U_0^2}{2g} \sin^2\alpha, \tag{17.16}$$

and the flight time is $T = L/U_0 \cos\alpha$.

We could show that for a specified value of launch velocity U_0, maximum range is obtained when the launch angle is

$$\alpha_m = \arctan \frac{U_0}{\sqrt{U_0^2 + 2gH}}, \tag{17.17}$$

where H may be positive or negative. When $H = 0$, this equation gives $\alpha_m = 45°$, as we obtained before. The maximum range is

$$L_m = \frac{U_0}{g} \sqrt{U_0^2 + 2gH}\,. \tag{17.18}$$

Shot-Putting a Shot

According to the numerical values listed in table 17.1, a shot-put shot has a drag-weight ratio $n = 0.005$. Since this is a very small number, air resistance can be neglected in the computation of the trajectory of the shot.

The origin of the trajectory is taken to be the outstretched hand of the shot-putter at the instant of release; his hand is 7.5 feet above the ground. Accordingly, as shown in figure 17.2(b), $H = -7.5$ ft at the impact point. Now suppose that our shot-putter decides on a launch velocity $U_0 = 45$ ft/s. Since he is not only an excellent athlete but also a very clever mathematician, he quickly calculates, from equation (17.7), that the launch angle giving maximum range is $\alpha_m = 41.9°$. The distance of the put, computed from equation (17.18), is $L_m = 70.0$ ft. For comparison, the world's record (in 1995) for the sixteen-pound shot is 75.9 feet.

From equations (17.16), the coordinates of the maximum point are $x_m = 31.2$ ft and $y_m = 14.0$ ft. These distances are measured from the origin, which is already 7.5 feet above ground level, so the maximum height of the shot above the ground is $7.5 + 14.0 = 21.5$ ft. The flight time of the shot is $T = 2.1$ s.

High Trajectories and Low Trajectories

Let us go back to the equation for the trajectory, equation (17.14), substitute $x = L$, $y = H$, and then solve it, not for L, but instead for $\tan \alpha$. For this computation we need the trigonometric relationship $\sec^2 \alpha = 1 + \tan^2 \alpha$. Some algebra gives us

$$\tan \alpha = \frac{U_0^2}{gL}\left[1 \pm \sqrt{1 - \frac{gL}{U_0^2}\left(\frac{gL}{U_0^2} + \frac{2H}{L}\right)}\right], \qquad (17.19)$$

where, as before, H may be positive or negative.

The plus-or-minus sign on the right-hand side of this equation indicates that, in general, there are two possible values of α for the trajectory connecting the launch point and the impact point: one is a high, peaked trajectory and the other is a low, flat trajectory. To illustrate this, we look at an example involving a basketball.

How to Make a Free Throw in Basketball

From table 17.1 we note that the drag-weight ratio of a basketball is $n = 0.085$. This is sufficiently small to allow us to neglect air resistance in the computation of the trajectory of a basketball, at least for low-velocity free throws.

The origin of the trajectory—the launch point—is located at the player's hand at the instant the ball is tossed. We assume this height to be $y_0 = 7.0$ ft. The hoop is at a height $y_h = 10.0$ ft above the floor. Accordingly, as indicated in figure 17.2(a), $H = y_h - y_0 = +3.0$ ft. The horizontal distance, measured along the floor, between the free-throw line and the centerline of the hoop is $L = 15$ ft.

So, we have the following numerical values: $U_0 = 25$ ft/s, $L = 15$ ft, and $H = +3$ ft. Substituting these numbers into equation (17.19) gives the two launch angles: $\alpha_1 = 59.4°$ and $\alpha_2 = 41.9°$.

PROBLEM 1. We determined that there are two possible values of the launch angle α for a free-throw shot. Show that the trajectory corresponding to the second value, $\alpha_2 = 41.9°$, is too flat; the basketball approaches the hoop at too sharp an angle to pass through the hoop. For your information, the diameter of the ball is $D = 9.6$ in. and the diameter of the hoop is $D_h = 17.7$ in.

PROBLEM 2. Show that the basketball reaches a maximum height of 14.2 feet above the floor.

PROBLEM 3. Confirm that the hang time of the free-throw shot is 1.2 seconds.

Water Jets and Water Fountains

For the remainder of this chapter we forget about sporting balls and projectiles. Instead, we are going to look into some aspects of water jets and water fountains. As before, we shall assume that the effects of air resistance can be ignored. In other words, we pretend that our water jet is in a vacuum.

With this assumption, it turns out that the equation for the trajectory of a water jet is the same as that for a projectile:

$$y = (\tan\alpha)x - \frac{g}{2U_0^2}(\sec^2\alpha)x^2, \tag{17.20}$$

where (x, y) are the coordinates of any point on the trajectory, α is the angle of the trajectory at the origin $(x = 0, y = 0)$, and U_0 is the jet velocity at the origin. In addition, all the numerous other equations we developed for sporting balls can also be applied to water jets.

The Fountains of the Het Loo Palace in the Netherlands

Without doubt, the most spectacular and beautiful application of water jets is their use in water fountains. We see these lovely artistic displays in many places: in parks and formal gardens and at shopping malls and commercial plazas. Assuredly, water fountains represent a delightful junction of mankind's talents in art and architecture, science, and technology.

Among the most beautiful and impressive water fountains in the world are those of the Het Loo Palace in Apeldoorn, Netherlands. As figure 17.3 shows, the artistic focal points of the fountains are two large spheres, each mounted on a pedestal. They are called the Celestial Sphere and the Terrestrial Globe. Each has a diameter of 1.80 meters; the top of each is 3.10 meters above ground. The spheres are precisely oriented to the latitude of Het Loo (52.2°N). Etched on the surface of the Celestial Sphere are the signs of the zodiac. The maps on the two spheres are based on those of the Italian cartographer Vincenzo Coronelli (1659–1718).

Jets of water are discharged from numerous small nozzles situated at various longitudes and latitudes on the surfaces of the spheres. The jets follow beautiful parabolic paths as they plunge into the pool below.

It would not be difficult to analyze the water jet hydrodynamics of the Het Loo spheres. To do so would require some analytic geometry and spherical trigonometry along with the jet trajectory mathematics. However, to simplify things, we replace the actual sphere (of 1.80 m diameter) with a so-called point source (of zero diameter). Then we pretend that water jets flow from the point source at specified values of launch angle α. We assume that the

FIG. 17.3

Fountain of the Celestial Sphere and (in the distance) Fountain of the Terrestrial Globe, Het Loo Palace. (Photograph provided by Rijksmuseum Paleis Het Loo.)

source is located at an elevation of 2.5 meters above the pool into which the water jets plunge. Accordingly, $H = -2.5$ m. The jet velocity is $U_0 = 5.0$ m/s. Also, recall that in the metric system of units, $g = 9.82$ m/s^2.

Next, using equation (17.20), jet trajectories are calculated for various values of α. The results are seen in figure 17.4 for values ranging from $\alpha = +90°$ (shooting straight upward) to $\alpha = -90°$ (shooting downward). From equations (17.17) and (17.18), we find that the launch angle $\alpha_m = 30.15°$ gives the maximum range, $L_m = 4.38$ m.

Observe in figure 17.4 that irrespective of the value of α, the jet trajectory never gets beyond the curve identified in the figure

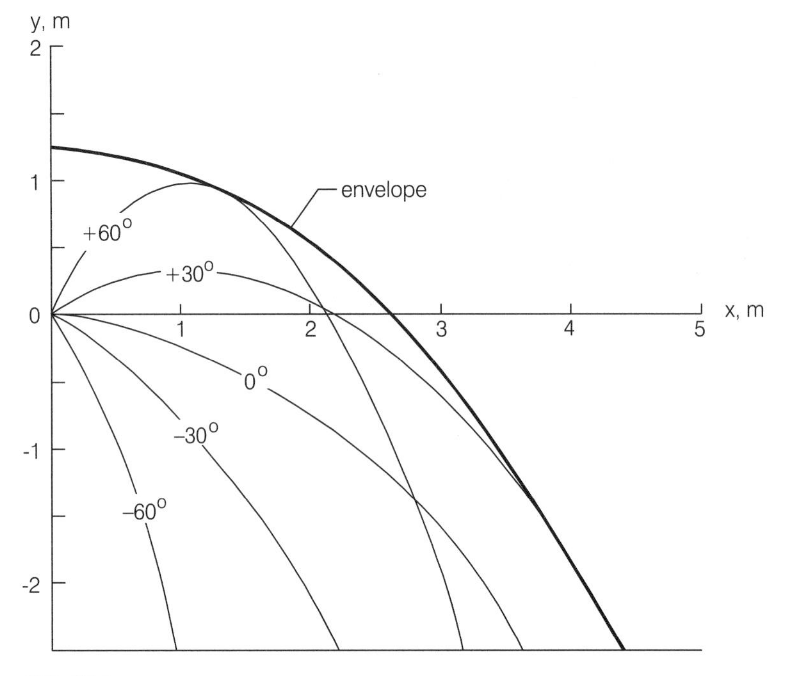

FIG. 17.4

Flow of water jets from a point source. The envelope defines a line or surface of safety.

as the "envelope." In ballistics, this boundary is also known as the "line of safety"; no projectile, launched with velocity U_0, can cross this line, regardless of the launch angle.

The equation of the envelope can be determined by using the following *rule*: The envelope of the family of curves $f(x, y, \alpha) = 0$ is given by the pair of equations

$$f(x, y, \alpha) = 0 \quad \text{and} \quad \frac{\partial f}{\partial \alpha}(x, y, \alpha) = 0. \tag{17.21}$$

Using equation (17.20) and applying this rule, we obtain the equation of the envelope,

$$y = \frac{U_0^2}{2g} - \frac{gx^2}{2U_0^2}. \tag{17.22}$$

This is the curve seen in figure 17.4. It is interesting that the envelope of all the parabolic trajectories is also a parabola. We could show that the focus of the envelope parabola is located at the origin $(0, 0)$. Clearly, when the envelope is rotated about the y-axis, we obtain a parabola of revolution. Thus the "surface of safety" is a paraboloid.

Hitting a Target with a Dwindling Water Jet

We shall conclude this chapter with some experimental work. Here is the problem. We are aiming a garden hose in such a way that the trajectory of the water jet strikes a small fixed target on a vertical flat surface some distance away. For some reason, intentional or otherwise, the velocity U_0 of the jet leaving the nozzle begins to dwindle (i.e., reduce). The problem: if we have a mathematical description of how U_0 changes with time—that is, $U_0 = f(t)$, then how should the launch angle α be changed with time, that is, $\alpha = f(t)$, in order to keep the impact point of the jet on the target?

For our experiments, we can probably devise apparatus more sophisticated than a garden hose nozzle; perhaps a hypodermic syringe or a toy water gun mounted on an axle would be better. In any event, for our experiments as shown in figure 17.5, we suppose that the target is located $L = 50$ cm horizontally from the nozzle and $H = -25$ cm vertically below it. Suppose also that the jet velocity, U_0, decreases linearly with time according to the expression

$$U_0 = U_* \left(1 - \frac{t}{t_*}\right). \tag{17.23}$$

This equation says that when time $t = 0$, then $U_0 = U_*$, and when $t = t_*$, then $U_0 = 0$. We shall take $U_* = 315$ cm/s and $t_* = 800$ s.

Next, using equation (17.19) and the numerical values indicated above, we compute the values of the launch angles α for various values of time, t. Thus, when $t = 0$, equation (17.23) gives

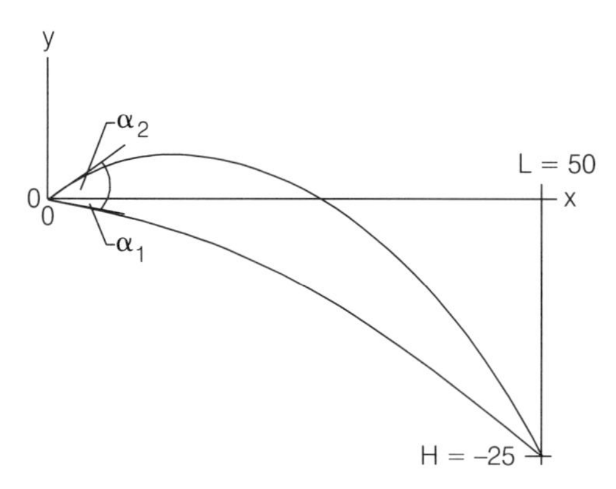

FIG. 17.5

Hitting a target with a dwindling water jet

$U_0 = 315$ cm/s; equation (17.9) yields the two values $\alpha_1 = -13.4°$ and $\alpha_2 = 76.9°$.

We see that the jet can hit the target with a low, flat trajectory or a high, peaked trajectory.

Similar calculations are made for increasing values of time. For example, when $t = 100$ s, equation (17.23) gives $U_0 = 275$, and equation (17.19) yields $\alpha_1 = -9.3°$ and $\alpha_2 = 72.8°$. Computed results are shown in figure 17.6, which indicates the two values of the launch angle that at time t will allow the jet to hit the target.

From equation (17.19) we establish that a critical time, t_c, occurs at 358 seconds. At that instant the jet velocity is $V_0 = 174$ cm/s and the two launch angles are both 31.7°. In other words, the two trajectories merge into a single trajectory. For times greater than 358 seconds, the jet can no longer reach the target. Our experiment has ended.

Trajectory Problems

PROBLEM 1. Shown in figure 17.7 is a children's swing such as one finds in the playground of a school or in a park. The length of the "pendulum" of the swing is R. A lively young lady, seated on the swing, is the prospective "projectile."

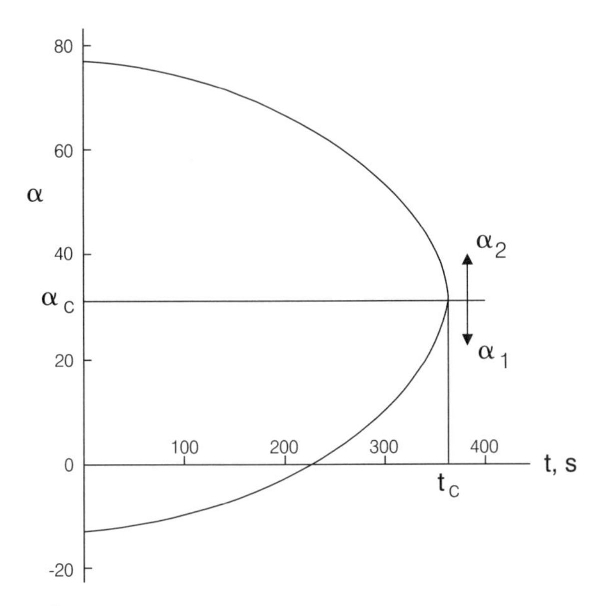

FIG. 17.6

Launch angles required to hit the target. $t_c = 358$ s, $\alpha_c = 31.7°$.

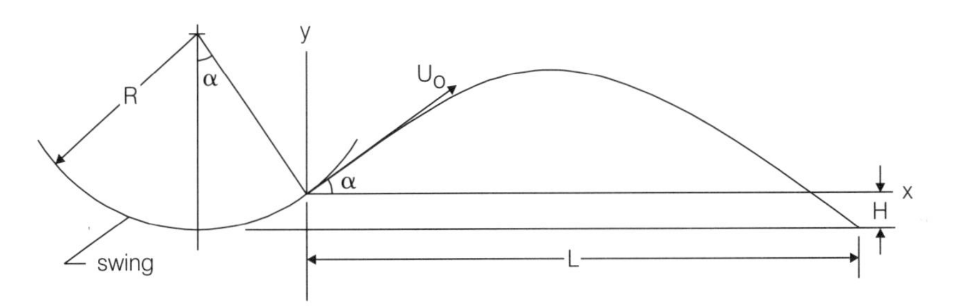

FIG. 17.7

Range of a projectile ejected from a swing

Here is the problem. The young lady (i.e., the projectile) decides to eject herself from the swing when the swing angle α is such that she attains a certain range L, as indicated in the figure. If $R = 5.0$ m and using $g = 9.82$ m/s^2, what is the value of the launch angle, α_m, corresponding to maximum range, L_m?

Hints. (*a*) Use equation (17.15) with negative value of H, (*b*) use trigonometry to get the equation $H = R(1 - \cos\alpha)$, and (*c*) use conservation of energy to obtain the relationship $U_0 = \sqrt{2g(R - H)}$. The answers are: $\alpha_m = 35.0°$, $L_m = 8.82$ m.

PROBLEM 2. The velocity U_0 of a water jet flowing from a small orifice in a tank is given by Torricelli's equation, $U_0 = C\sqrt{2gh}$, where C is a so-called contraction coefficient and h is the depth of water above the orifice. Verify that a *cylindrical* tank, as it drains through the orifice, produces the *linear* relationship of equation (17.23).

PROBLEM 3. Show that t_*, defined in equation (17.23), is given by the expression $t_* = (A/aC)\sqrt{2h_*/g}$, in which A is the cross-sectional area of the tank, a is the area of the orifice, and h_* is the initial depth.

There are numerous books that deal with applications of mathematics in sports. Three very good ones are de Mestre (1990), Hart and Croft (1988), and Townend (1984).

18

Balls and Strikes and Home Runs

In our study of trajectories in the preceding chapter, it was established that when an object, for example, a tennis ball, is moving through a vacuum (i.e., no air), it is not difficult to determine its path. The problem is easy because there is only one force acting on the object: gravity force. In this case, the trajectory is simply a parabolic curve. However, when the object is moving through the air, it is usually necessary to consider also the effect of another force: drag force due to the air. Generally, this makes computation of the trajectory much more difficult.

To help us decide when the effects of air resistance can be neglected in trajectory analysis, recall that we constructed a parameter, n, which expresses the ratio of drag force to gravity force. We then computed values of this drag-weight ratio for several kinds of sporting balls. These values are listed in table 17.1. When this ratio, n, is numerically small (in comparison with 1.0), the drag force due to air can be neglected in trajectory analysis. However, when the n ratio is large, air resistance cannot be ignored.

For example, from table 17.1, we conclude that air resistance can be neglected if we are computing the flight path of a shot put ($n = 0.005$) or a basketball, at least for a relatively low-velocity free throw ($n = 0.085$). However, we cannot ignore the drag force in trajectory analyses of baseballs ($n = 1.01$), tennis balls ($n = 2.19$), or golf balls ($n = 2.83$).

We start our chapter with an analysis of the problem of baseball trajectory determination. At the outset, we list some of the physical characteristics of baseballs.

circumference, C: 9.0 to 9.5 inches
diameter, D: 2.90 inches, 0.241 feet, 0.0736 meters
weight, W: 5.12 ounces, 0.32 pounds, 0.145 kilograms
density, ρ: 1.356 slugs/ft^3, 694 kg/m^3
roughness elements: 216 stitches of red cotton yarn along a 16-inch seam on a cowhide cover

Flight of a Baseball: Pitcher's Mound to Home Plate

Before we start our basic analysis, let's examine a fairly easy problem that does include the drag force. The problem is how long does it take for a pitched baseball to travel from the pitcher's mound to home plate? It is assumed that the pitch is in the horizontal (x) direction and that there is no spin on the ball. Also, as in all the analyses of this chapter, we assume there is no wind.

As we have done repeatedly in earlier chapters, we employ Newton's second law of motion to obtain the answer to our problem. After the pitcher releases the ball, the only force in the horizontal direction is the drag force, $F_D = (1/2)\rho_a C_D A U^2$. So we have

$$-\frac{1}{2}\rho_a C_D A U^2 = m\frac{dU}{dt}, \tag{18.1}$$

where ρ_a is the density of air, C_D is the drag coefficient, A is the projected area of the spherical ball, $a = dU/dt$ is the ball's acceleration, and $U = dx/dt$ is its instantaneous velocity. In these relationships, t is time and x is distance measured from the pitcher's mound ($x = 0$) toward home plate. Accordingly,

$$-\frac{1}{2}\rho_a C_D \frac{\pi}{4} D^2 U^2 = \rho \frac{\pi}{6} D^3 U \frac{dU}{dx}, \tag{18.2}$$

in which ρ is the density and D the diameter of the ball. The numerical value of the drag coefficient C_D depends on several things; we will come back to this in a moment.

Simplifying equation (18.2) gives

$$-\frac{3\rho_a C_D}{4\rho D}\int_0^x dx = \int_{U_0}^{U}\frac{dU}{U}. \qquad (18.3)$$

The lower limits on the integrals stipulate that the velocity of the ball is U_0 the instant it leaves the pitcher's hand at $x = 0$. Integrating,

$$U = U_0 e^{-kx}, \qquad (18.4)$$

where $k = 3\rho_a C_D/4\rho D$. This is an interesting result. It says that the velocity of the baseball decreases with distance like a negative exponential.

Let us use some numbers; the distance between the pitcher's mound and home plate is L. We have $\rho_a = 0.0024$ slug/ft^3, $C_D = 0.30$, $D = 0.241$ ft, $\rho = 1.356$ slug/ft^3, $k = 0.00165$ 1/ft, $L = 60.5$ ft, and $U_0 = 125$ ft/s (85 mi/hr). Substituting these values into equation (18.4), with $x = L$, gives $U_L = 113$ ft/s (77 mi/hr). This is the velocity of the ball as it crosses home plate. We note that air resistance reduces the ball's velocity by about 10%.

To complete our analysis of the problem, we use the equation $U = dx/dt$. From equation (18.4), the following expression is obtained:

$$\int_0^L e^{kx}\,dx = U_0\int_0^T dt, \qquad (18.5)$$

where T is the flight time of the ball. Integrating and solving for T gives the result

$$T = \frac{1}{kU_0}(e^{kL} - 1). \qquad (18.6)$$

Employing the same numbers as before gives $T = 0.51$ s. This is the time required for the ball to travel from the pitcher's mound to home plate. We make two observations:

1. The batter has about half a second to decide whether he should swing at the ball and, if he does, to carry out the actual swing.
2. Even though the trajectory of the ball is primarily in the horizontal (x) direction, gravity force is still acting on the ball in the vertical (y) direction. During the flight time, $T = 0.51$ s, the ball falls a distance $y = (1/2)gT^2 = (1/2)(32.2)(0.51)^2 = 4.2$ ft. Very interesting.

The Drag Force on Sporting Balls

As we have seen, the drag force on a sphere is given by the expression

$$F_D = \frac{1}{2}\rho_a C_D \frac{\pi}{4} D^2 U^2, \tag{18.7}$$

in which ρ_a is the density of air, D is the diameter of the sphere, U is its velocity, and C_D is the drag coefficient. It turns out that C_D depends on the numerical value of the so-called Reynolds number, $Re = \rho_a UD/\mu_a$, where μ_a is the viscosity of air. In addition, C_D depends on the relative roughness of the sphere, ϵ/D, in which ϵ is the roughness height. The roughness caused by the seam and stitches on a baseball, or the dimples on a golf ball, or the fuzz and groove on a tennis ball determines the numerical value of ϵ.

If the ball is spinning, as is almost always the case with baseballs and golf balls, the drag coefficient also depends, to a limited extent, on the magnitude of the spin parameter, $\omega R/U$, where ω is the angular velocity of the spinning sphere, measured in radians per second, and R is the radius of the ball.

The plot of figure 18.1, from Bearman and Harvey (1976), shows the dependence of the drag coefficient of a sphere, C_D, on the Reynolds number, *Re*, and the relative roughness, ϵ/D. In the present analysis, we neglect the minor effect of the spin parameter on the drag coefficient.

An interesting and important feature is revealed in figure 18.1. Suppose the magnitude of the Reynolds number is $Re = 187{,}000$; this is the value of *Re* in our pitched-ball example above, assuming an air temperature of 68°F. Now if the sphere is completely smooth ($\epsilon = 0$), the value of the drag coefficient is about $C_D = 0.5$. However, if the sphere is rough ($\epsilon > 0$), as in the case of a baseball or golf ball, the magnitude of the drag coefficient is much less, $C_D = 0.1$ to 0.3. The substantially reduced drag coefficient of a rough sphere means that the drag force is considerably less, and consequently there is less energy loss due to air resistance during the ball's flight.

The outcome of all this is that a rough sphere is able to travel considerably farther than a smooth sphere. This fact was discov-

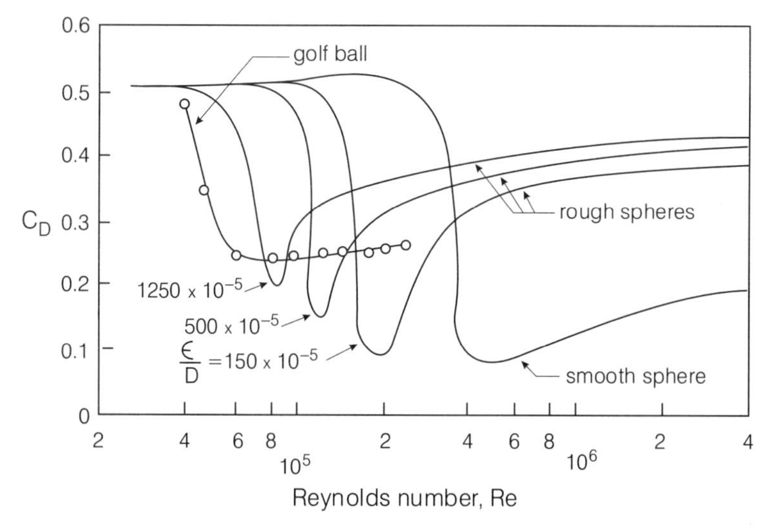

FIG. 18.1

Drag coefficient curves for smooth and rough spheres. (From Bearman and Harvey 1976.)

ered a long time ago by the early developers and designers of baseballs and golf balls. By the way, if you are interested in detailed information about the aerodynamics of flow about smooth and rough spheres, two excellent references are suggested: Adair (1990) and Watts and Bahill (1990). In addition, Mehta (1985) presents a comprehensive review of the aerodynamics of sports balls.

The Lift Force on Sporting Balls

So you want to make our problem even more complicated than it is already? Not difficult to do. It turns out that not only is there a *drag force* acting on the sphere, there is also, if the sphere is spinning, a *lift force*. Such a force can create much joy (increased range of the ball) or much sorrow (foul balls, pop flys, hooks, and slices).

Now examine figure 18.2. This shows a moving sphere rotating counterclockwise with angular velocity ω. The velocity of the sphere itself, from left to right, is U. In this case, we have *backspin*. Now pretend that the spinning sphcrc is stationary and that the air is moving, from right to left, with velocity U. If we assume a "no-slip" condition concerning the velocity of the air

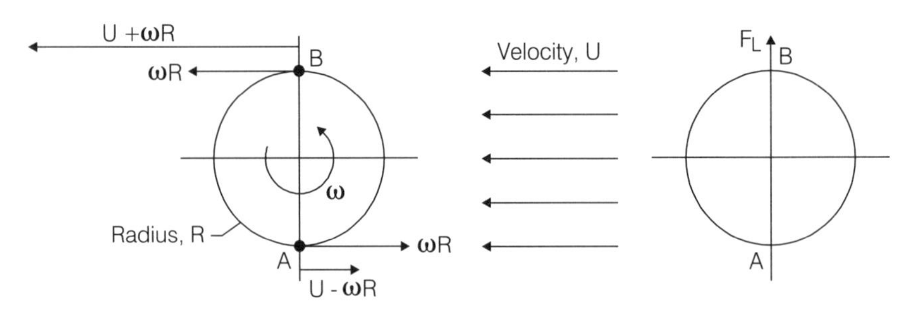

FIG. 18.2

Definition of the Magnus force. A sphere of radius R spins with angular velocity ω in an air flow of velocity U, creating a lift force F_L the Magnus force.

and that of an adjacent boundary, then the velocities at points A and B are, respectively, $U_A = U - \omega R$ and $U_B = U + \omega R$.

The very important and well-known Bernoulli equation relates the velocity of a fluid at a particular point to the pressure at that point. That is,

$$p_A + \tfrac{1}{2}\rho_a U_A^2 = p_B + \tfrac{1}{2}\rho_a U_B^2, \tag{18.8}$$

where p_A and p_B are the pressure at points A and B. Now since velocity U_A is less than velocity U_B, it is clear from the Bernoulli equation that the pressure p_A is greater than the pressure p_B. Consequently, a net force acts in the direction from A to B perpendicular to the trajectory of the sphere. This force has a special name, the Magnus force, named after the German engineer who investigated this phenomenon around 1850. Clearly, the Magnus effect creates a lift force, F_L, on the sphere.

As we did in the case of drag force, we write an equation expressing the lift force on a sphere

$$F_L = \frac{1}{2}\rho_a C_L \frac{\pi}{4} D^2 U^2, \tag{18.9}$$

where C_L is the lift coefficient.

Experimental studies carried out by numerous investigators are reported by Watts and Bahill (1990) indicate that the lift coefficient of a spinning sphere, C_L, is numerically equal to the spin parameter, $\omega R/U$, for values of the spin parameter less than about 0.4. That is,

$$C_L = \frac{\omega R}{U}. \tag{18.10}$$

Again, let's use some numbers. The radius of a baseball is $R = 0.121$ ft. Suppose that the pitcher throws the ball at a velocity $U = 125$ ft/s, with a backspin of $N = 1{,}500$ rev/min. Then $\omega = 2\pi N/60 = 2\pi(1{,}500)/60 = 157$ rad/s, and so $\omega R = 19.0$ ft/s. Accordingly, the spin parameter $\omega R/U = (19.0)/125 = 0.15$, and so, from equation (18.10), we obtain $C_L = 0.15$.

Finally, substituting equation (18.10) into equation (18.9) gives

$$F_L = \frac{\pi}{16}\rho_a \omega D^3 U, \quad (18.11)$$

which says that the lift force, F_L, is proportional to the first power of the velocity, U. This result has a measure of confirmation in the well-known relationship in theoretical aerodynamics called the Kutta-Joukowski equation. This equation says that the lift force on an object in a fluid flow (e.g., an airfoil) is

$$F_L = \rho_a \Gamma U, \quad (18.12)$$

where ρ_a is the fluid density, U is the velocity of the object, and Γ is the so-called circulation about the object. Comparing equations (18.11) and (18.12) yields $\Gamma = (\pi/16)\omega D^3$ for a spinning sphere. The important result is that the circulation Γ is directly proportional to the spin velocity, ω; this seems logical.

The Trajectory Equations of Baseballs and Golf Balls

We are now ready to write down the equations we need to compute the trajectory of a spinning rough sphere moving through air. Once again we start with Newton's second law of motion. This time we use this relationship twice, in the horizontal (x) and the vertical (y) directions. A definition sketch is given in figure 18.3. This sketch shows the weight (W), drag force (F_D), and lift force (F_L) acting on the sphere at a particular point along the trajectory. The velocity of the sphere is U; its components in the horizontal (x) and vertical (y) directions are u and v, respectively; clearly, $U = \sqrt{u^2 + v^2}$. The angle between the tangent to the trajectory and the horizontal is θ. We note that $\cos\theta = u/U$ and $\sin\theta = v/U$.

Substituting the various indicated force components into the two equations of motion yields the following expressions for the trajectory with m the mass of the sphere and $W = mg$ its weight:

$$mu\frac{du}{dx} = -F_L \sin\theta - F_D \cos\theta, \quad (18.13)$$

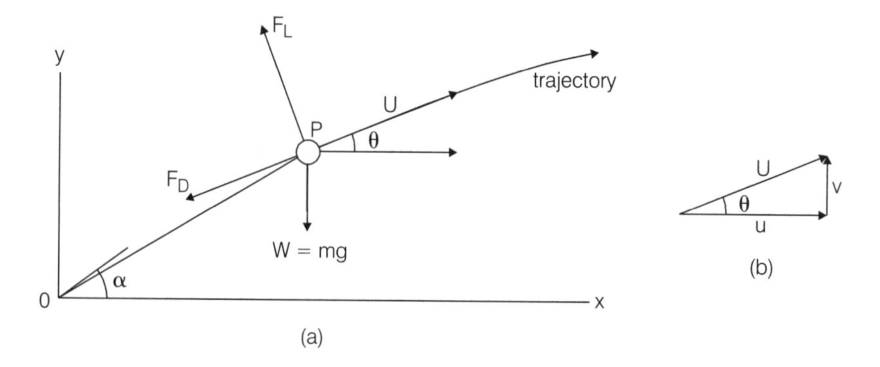

FIG. 18.3

Definition sketches showing (*a*) the forces acting on a sphere and the trajectory of the sphere and (*b*) the velocity components, *u* and *v*, in the *x* and *y* directions

and

$$mv\frac{dv}{dy} = F_L \cos\theta - F_D \sin\theta - mg, \tag{18.14}$$

where, again, $\cos\theta = u/U$ and $\sin\theta = v/U$.

If the sphere is moving in a vacuum (i.e., no air), we can set $\rho_a = 0$ and hence $F_L = 0$ and $F_D = 0$. In this case, as you might want to show, equations (18.13) and (18.14) reduce to the expressions yielding the parabolic trajectories of the previous chapter.

Effect of Spin on the Curve of a Baseball

The expressions for computation of baseball and golf ball trajectories given by equations (18.13) and (18.14) are called nonlinear first-order differential equations. Exact solutions are impossible to obtain. Consequently, it is necessary to solve them by numerical computation methods. We shall get to that shortly.

However, first let us continue with the pitched-ball problem we considered above. This time the ball has a backspin of ω radians per second about a horizontal axis perpendicular to the flight path. Consequently, there is a lift force on the ball acting upward.

We assume that the horizontal velocity, u, is very much larger than the vertical velocity, v. Accordingly, we take $\cos\theta = u/U = 1$ and $\sin\theta = v/U = 0$. In this case, equation (18.13) becomes

$$mU\frac{dU}{dx} = -F_D. \tag{18.15}$$

This is the same as Equation (18.2). Recall that the ensuing analysis of that expression gave us a ball flight time $T = 0.51$ s, between the pitcher's mound and home plate, and a vertical fall of $y = 4.2$ ft as the ball crosses home plate.

Now, with $\cos\theta = 1$ and $\sin\theta = 0$, equation (18.14) becomes

$$mv\frac{dv}{dy} = F_L - mg. \tag{18.16}$$

Utilizing equation (18.11), this expression becomes

$$v\frac{dv}{dy} = \frac{3\rho_a\omega}{8\rho}U - g. \tag{18.17}$$

We let

$$g_* = \frac{3\rho_a\omega U}{8\rho} - g, \tag{18.18}$$

in which g_* is a kind of modified gravitational acceleration due to the Magnus force. For the horizontal velocity U, we can use the average velocity, $U = (U_0 + U_L)/2$. So equation (18.17) becomes

$$\frac{dv}{dt} = g_*, \tag{18.19}$$

using $v = dy/dt$. From this equation, we easily determine that $v = g_* t$ and $y = (1/2)g_* t^2$. Consequently, during the flight time T, the spinning ball falls a distance $y = (1/2)g_* T^2$. Using the previous numbers, we get $g_* = 19.8\ \text{ft/s}^2$ (instead of $32.2\ \text{ft/s}^2$) and so $y = 2.6$ ft.

In our earlier computation involving a nonspinning baseball, we calculated a fall distance of 4.2 feet. So it turns out that the Magnus force acting vertically upward causes the ball to cross home plate a distance 1.6 feet above where it would have been had there been no spin. The pitcher, of course, hopes that the batter does not anticipate this.

Lateral Curves of Baseballs

The numerical example we just completed involved a "rising fastball": the pitcher gives a backspin to the ball to cause it to curve upward in the vertical plane. Alternatively, the pitcher may decide to throw a curve ball in the horizontal plane so that the ball will pass to the left or to the right as it crosses home plate. In this case, the equation of motion is

$$mw\frac{dw}{dz} = F_L, \qquad (18.20)$$

in which z is the direction perpendicular to the vertical (x, y) plane and w is the velocity component of the spinning ball in the z direction. If the ball spins counterclockwise about the vertical axis (looking downward), the ball will curve to the left; if it spins clockwise, it will curve to the right.

PROBLEM. Using the above information, you should be able to show that the amount of the ball's deflection, d, as it crosses home plate—that is, the lateral curve—is given by the equation

$$d = \frac{3\rho_a \omega U}{16\rho}T^2. \qquad (18.21)$$

There Are All Kinds of Pitches

Many volumes have been written about the art and science of pitching baseballs and, in the years to come, assuredly many more volumes will appear. For our purpose, here is simply a short

TABLE 18.1

Typical characteristics of various types of pitches by major league pitchers

Type of pitch	*Velocity U, mi/hr and ft/s*	*Typical spin rate N, rpm*	*Rotations en route to plate*
Fastball	85 to 95 125 to 139	1,600	11
Slider	75 to 85 110 to 125	1,700	13
Curve ball	70 to 80 103 to 117	1,900	16
Change-up	60 to 70 88 to 103	1,500	14
Knuckle ball	60 to 70 88 to 103	25 to 50	1/4 to 1/2

Source: Data from Watts and Bahill (1990).

list of some of the many kinds of pitches:

rising fast ball	sinker
sinking fast ball	screwball
curve ball	fork ball
slider	knuckle ball

The type of pitch depends on the orientation of the pitcher's grip with respect to the seam of the ball and on the amount and type of spin the pitcher decides to give. Concise descriptions of many types of pitches and their characteristics are given by Watts and Bahill (1990), from which reference table 18.1 is adapted.

How to Hit a Home Run

We have completed our brief examination of the topic of *pitching* baseballs; we move on to the topic of *hitting* them. The crucial topic of instantaneous interaction between the ball and the bat and the associated interesting problems of collision dynamics will be bypassed. These problems are considered at length by Adair (1990) and Watts and Bahill (1990).

Look at equations (18.13) and (18.14). These are the relationships that enable us to compute the trajectory of a batted baseball. Earlier, it was pointed out that it is impossible to obtain exact solutions to these equations. Consequently, it is necessary to resort to numerical methods to obtain approximate answers. A reliable and relatively easy method to solve these nonlinear first-order differential equations numerically is the so-called Runge–Kutta method. Briefly, here is the way it works.

We presume we know the initial velocity, U_0, and the launch angle, α, at the origin, $x = 0, y = 0$. Divide the two equations by the mass, m, and write the acceleration components in the form $a_x = du/dt$ and $a_y = dv/dt$. Now with the initial velocity components u_0 and v_0 known, it is easy to compute all the terms on the right-hand side of the equations. So we can easily calculate the initial acceleration components, a_{x0} and a_{y0}.

By definition, $a_{x0} = (u_1 - u_0)/\Delta t$ and, by the same token, $a_{y0} = (v_1 - v_0)/\Delta t$, where Δt is a small time increment (e.g., $\Delta t = 0.03$ s). From these relationships, we compute the new velocity components, u_1 and v_1. Furthermore, during the Δt time interval, the ball has moved a distance

$$\Delta x = (1/2)(u_1 + u_0)\,\Delta t \qquad \text{and}$$

$$\Delta y = (1/2)(v_1 + v_0)\,\Delta t.$$

We repeat this calculation procedure and continue doing so until the entire trajectory is covered. This method of solution is easy to handle with a computer.

Baseball trajectory computations of this type were carried out by Watts and Bahill (1990); the results are seen in figure 18.4. For these calculations, the initial velocity is $U_0 = 110$ mi/hr = 161 ft/s and the launch angle is $\alpha = 30°$. The dashed curve designated "no drag" is the parabolic curve corresponding to the trajectory in a vacuum considered in the previous chapter. In table 18.2, a listing is presented of the range (L), maxi-

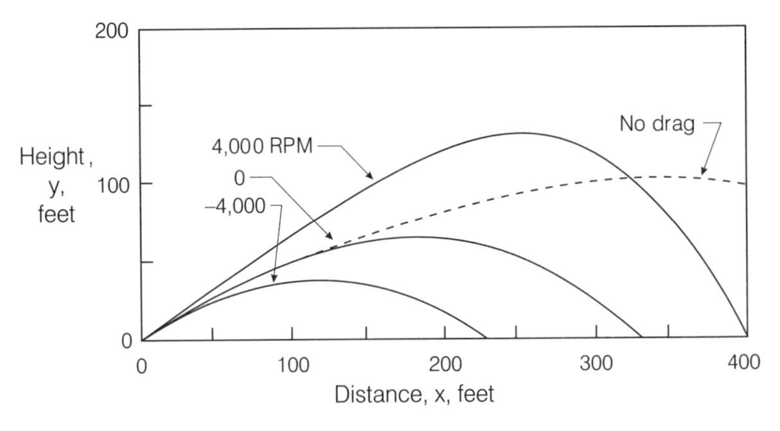

FIG. 18.4

Trajectories of batted baseballs computed by the Runge–Kutta method. Initial velocity $U_0 = 110$ mi / hr, launch angle $\alpha = 30°$. (From Watts and Bahill 1990.)

TABLE 18.2

Results of Runge–Kutta analysis of baseball trajectories with $V_0 = 110$ mi / hr, $\alpha = 30°$

Trajectory	*Range, L, ft*	*Maximum height, y_m, ft*	*Flight time, T, s*
No drag	697	101	5.0
4,000 rpm	400	137	3.5
0 rpm	329	69	2.7
−4,000 rpm	226	37	1.7

Source: Data from Watts and Bahill (1990).

mum height (y_m) and flight time (T) of the trajectories shown in figure 18.4.

A Thrilling Baseball Game

It's the bottom of the ninth with two outs. Our friend Joe Smog is at bat. On a full count, Joe hits the ball with an initial velocity

$U_0 = 110$ mi/hr $= 161$ ft/s, at a launch angle $\alpha = 30°$, and with a backspin $N = 4{,}000$ rpm (419 rad/s). The ball sails through the air for $T = 3.5$ s, reaches a maximum height $y_m = 137$ ft and travels a distance $L = 400$ ft.

The ball just clears the center field fence so Joe gets a home run. He is really happy because this was his first time at bat after being called up from the minors, and also the bases were loaded.

19

Hooks and Slices and Holes in One

We have completed our analysis of baseball trajectories and move on to an examination of the trajectories of golf balls. In many ways the two studies are similar. As an introduction to our analysis, we list some of the characteristics of golf balls:

circumference, C: 5.10 to 5.28 inches
diameter, D: 1.65 inches, 0.138 feet, 0.042 meters
weight, W: 1.62 ounces, 0.101 pounds, 0.046 kilograms
density, ρ: 2.296 slugs/ft^3, 1186 kg/m^3
roughness elements: 330 to 400 small dimples spaced uniformly over the ball's surface

The Trajectory Equations

In our analysis of the flight path of a baseball, recall that we derived the equations of motion in terms of a cartesian or rectangular (x, y) coordinate system, as shown in figure 18.3. As a result, we obtained equations (18.13) and (18.14).

This time, for a change, we use a polar or so-called intrinsic (s, n) coordinate system. Figure 18.3 again serves as the definition sketch; the center of the moving sphere is the (s, n) origin. The s direction corresponds to the instantaneous direction of the ball at point P; the n direction is perpendicular to the s direction at P.

The drag force F_D acts along s, the lift force F_L acts along n, and, as always, the weight W acts vertically downward.

To obtain the trajectory equations, we again employ Newton's equation of motion twice, first in the s direction and then in the n direction. In the s direction, the acceleration is given by $a_s = dU/dt$. In the n direction, the acceleration is $a_n = U^2/r$, where r is the radius of curvature of the trajectory at point P. This analysis yields the following equations:

$$m\frac{dU}{dt} = -F_D - mg\sin\theta, \tag{19.1}$$

and

$$mU\frac{d\theta}{dt} = F_L - mg\cos\theta, \tag{19.2}$$

in which θ is the angle between the trajectory at point P and the horizontal. The drag force F_D is

$$F_D = \frac{1}{2}\rho_a C_D \frac{\pi}{4} D^2 U^2 = k_1 U^2, \tag{19.3}$$

and the lift force F_L is

$$F_L = \frac{\pi}{16}\rho_a \omega D^3 U = k_2 U. \tag{19.4}$$

These two equations define the constants k_1 and k_2, which we shall be using shortly. This approach follows that of Hart and Croft (1988).

An Example of a Golf Ball Trajectory

In our numerical examples involving baseball trajectories, we employed the English or engineering system of units (foot, pound, second). This time, for the golf ball problem, let's use the SI or metric system (meter, kilogram, second).

Here is an example of the type of problem we want to solve:

PROBLEM. Determine the trajectory of a golf ball having an initial velocity $U_0 = 60$ m/s, a launch angle $\alpha = 12°$, and a backspin $N =$ 4,000 rev/min. Ambient temperature is 20°C; sea-level pressure.

First, we compute the Reynolds number and the spin parameter. The density of air at 20°C is $\rho_a = 1.205$ kg/m^3 and the viscosity is $\mu_a = 1.81 \times 10^{-5}$ kg/(m/s). Accordingly, the Reynolds number is

$$Re = \frac{\rho_a UD}{\mu_a} = \frac{(1.205)(60)(0.042)}{1.81 \times 10^{-5}} = 1.68 \times 10^5.$$

From figure 18.1, we get the approximate value of the drag coefficient, $C_D = 0.25$. The angular velocity of the backspin, in radians per second, is $\omega = 2\pi(4{,}000)/60 = 419$ rad/s. Consequently, the spin parameter is

$$\frac{\omega R}{U_0} = \frac{(419)(0.042/2)}{60} = 0.15.$$

From equation (18.10), we obtain the value of the lift coefficient, $C_L = 0.15$.

As in the case of baseball trajectory analysis, it is not possible to obtain exact solutions to the golf ball trajectory equations, equations (19.1) and (19.2). Therefore, numerical analysis is necessary to obtain even approximate solutions. Hart and Croft (1988) employed the Runge–Kutta method to solve our numerical example. Their results are shown in table 19.1.

An Approximate Solution to the Trajectory Equations

If the assumption is made that the trajectory of a golf ball flight is very flat, that is, the angle θ is small, it is possible to obtain an approximate solution to the trajectory equations. Indeed, a remarkable sequence of short papers written many years ago by Peter Tait (1831–1901), a Scottish mathematical physicist,

TABLE 19.1

Results of Runge – Kutta numerical computation of golf ball trajectory, with $U_0 = 60$ m / s, $\alpha = 12°$, $N = 4{,}000$ rpm

t, s	*x, m*	*y, m*
1	47.7	8.8
2	81.7	11.0
3	108.7	7.3
4	131.4	−1.7

Range $L = 128.2$ m

time of flight $T = 3.8$ s

Maximum height of ball, $y_m = 11.0$ m at $x_m = 81.7$ m

Source: Data from Hart and Croft (1988).

provides a simple and surprisingly accurate method for solving the golf ball trajectory problem. Tait must have been a remarkably interesting person. After several years at Cambridge University in England, he returned to his native Scotland and spent over forty years as a professor of physics and mathematics at Edinburgh University. He was a contemporary of Lord Kelvin and Clerk Maxwell.

There is no doubt that Tait's favorite pastime was golf; he was an accomplished player. He spent most of his holidays at nearby Saint Andrews and he played the famous golf course there a great many times. He invented instruments to measure ball velocities, flight angles, and ball heights. He even devised a fluorescent golf ball so that he and his friends could play the game quite late on those dark winter afternoons of north Scotland.

Tait published the remarkable sequence of papers on the physics of golf in *Nature* magazine: Tait (1890, 1891, 1893). Another lengthy nonmathematical article concerning golf appeared in the March 1896 issue of *Badminton Magazine*. Entitled "Long Driving," this publication is reprinted in the Tait memoirs authored by Knott (1911); perhaps this charming and very informative paper should be read by all serious golfers. A discussion of Tait's analysis of golf ball trajectories is given by de Mestre (1990).

Starting with equations (19.1) and (19.2) and making the assumption that the trajectory is very flat (i.e., sin $\theta = 0$, cos $\theta = 1$), Tait obtained the following equation for the trajectory:

$$y = \alpha x + C_1\left(e^{k_1 x/m} - 1 - \frac{k_1 x}{m}\right) - C_2\left(e^{2k_1 x/m} - 1 - \frac{2k_1 x}{m}\right), \tag{19.5}$$

in which

$$C_1 \frac{k_2 m}{k_1^2 U_0} \quad \text{and} \quad C_2 = \frac{gm^2}{4k_1^2 U_0^2}. \tag{19.6}$$

Here m is the mass of the golf ball, U_0 is its initial velocity, α is the launch angle, and g is the gravitational acceleration. The quantities k_1 and k_2 are defined by equations (19.3) and (19.4).

The time of flight T is given by the equation

$$T = \frac{m}{k_1 U_0}(e^{k_1 L/m} - 1). \tag{19.7}$$

We continue with the numerical example, presented by Hart and Croft, in which $U_0 = 60$ m/s, $\alpha = 12°$, and $N = 4{,}000$ rpm. In addition, Hart and Croft specify the following numerical values with SI units: $k_1 = 3.589 \times 10^{-4}$ and $k_2 = 5.834 \times 10^{-3}$. Substituting these numbers into equations (19.5) and (19.6) and carrying out the indicated calculations, we get the trajectory shown in figure 19.1. Note that the vertical scale of this figure has been exaggerated four times.

The simplicity of equation (19.5) makes it very easy to handle. The range, $x = L$, is determined by setting $y = 0$. A trial-and-error solution provides the result: $L = 133$ m. From equation (19.7), we calculate the time of flight, $T = 3.9$ s.

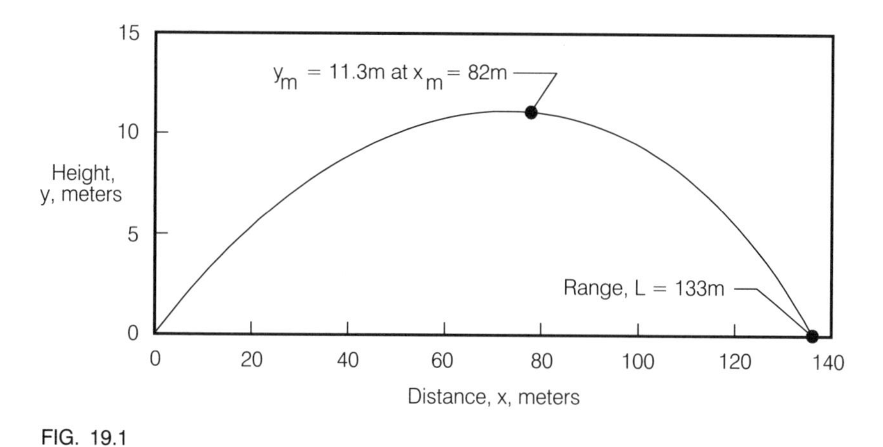

FIG. 19.1

Trajectory of a golf ball, computed from approximate solution by Tait. Initial velocity $U_0 = 60$ m / s, launch angle $\alpha = 12°$, and backspin $N = 4{,}000$ rpm.

Using the methods of differential calculus, it is not difficult to determine the maximum height of the ball. The result: $y_m = 11.3$ m when $x_m = 82$ m. We note that the answers provided by this approximate solution, equation (19.5), are in remarkably close agreement with the results obtained from the Runge–Kutta computations.

In his many publications about golf, Tait put much emphasis on the importance of spin and Magnus force for the behavior of the ball. With reference to equation (19.5), we note that the coefficient C_1 may be positive, zero, or negative, depending on whether the ball has backspin, no spin, or topspin.

All too frequently, the effects of spin are detrimental. If the spin axis is not horizontal, the ball may hook (curve to the left) or it may slice (curve to the right). Topspin about a horizontal axis exerts a negative lift force, which, acting in the same direction as gravity, hastens the ball's plunge to the ground. On the other hand, the effects of spin may be beneficial. Backspin about a horizontal axis exerts a positive lift force, which, acting vertically upward, increases the ball's altitude and extends its time of flight and range.

Returning to our numerical example, table 19.2 summarizes the effects of spin on the main parameters of the trajectory. We note that the effects are appreciable.

We can use equation (19.5) not only to determine the maximum height of the ball but also to find out if the trajectory has an inflection point. That is, is there so much backspin and lift on the ball that initially the ball curves upward? Here is how the analysis goes.

You will easily obtain the following equation for the *slope* of the trajectory by differentiating equation (19.5):

$$\frac{dy}{dx} = \alpha + C_1\frac{k_1}{m}(e^{k_1x/m} - 1) - C_2\frac{2k_1}{m}(e^{k_1x/m} - 1). \tag{19.8}$$

In turn, by differentiating this expression, we can get an equation for the *curvature*. Setting this equation equal to zero determines the location of the inflection point of the trajectory. The result is

$$x_c = \frac{m}{k_1}\log_e\left(\frac{k_2U_0}{gm}\right), \tag{19.9}$$

where x_c is the distance from the tee ($x = 0$). We note from this result that in order to have an inflection point, it is necessary that the product k_2U_0 be greater than the product gm. In our numerical example this is not the case. However, you may want

TABLE 19.2

Effects of spin on trajectory, with $U_0 = 60$ m/s, $\alpha = 12°$

Parameter	*Backspin* $N = +4{,}000$	*No spin* $N = 0$	*Topspin* $N = -4{,}000$
Range L, m	133	88	60
Maximum point			
x_m, m	82	50	35
y_m, m	11.3	5.9	3.9
Flight time T, s	3.9	2.1	1.3

to convince yourself that if the backspin were doubled to $N =$ 8,000 rpm, and everything else remained the same, there would be an inflection point located at $x_c = 56$ m, $y_c = 13.6$ m. This means that for values of x between 0 and x_c the trajectory is concave upward; for x larger than x_c it is concave downward.

How Is the Air Up There?

Before we leave the subject of sporting ball trajectories, a final question is raised: what is the effect of altitude on trajectory characteristics? That is, does the reduced density of air at higher elevations significantly increase the range, maximum height, and time of flight of a ball?

To provide a rational basis for answering this question, we combine the so-called general gas law with the equation for fluid statics to obtain the relationship

$$\rho = \rho_0 e^{-(mg/R_* T)z}, \tag{19.10}$$

in which ρ_0 is the density of air at sea level ($z = 0$) and ρ is the density at elevation z. The other quantities are $m = 28.96 \times 10^{-3}$ kg/mol, $R_* = 8.314$ joules/°K mol, $g = 9.82$ m/s^2, $\rho_0 = 1.205$ kg/m^3, and $T = $ °K = °C + 273°. If $T = 20$°C, equation (19.10) becomes

$$\rho = 1.205 e^{-0.000117z}. \tag{19.11}$$

For example, the elevation of Denver, the "mile-high city," is $z = 1{,}610$ m. Substituting this into equation (19.11) gives the result $\rho = 0.998$ kg/m^3. So the density of air at the altitude of Denver is only 83% of the density at sea level. Recall from our analysis that both the drag force and the lift force are directly proportional to the air density.

What is the consequence of all this? Does this give the Colorado Rockies baseball team a significant advantage over the other teams? If you play golf at the top of Pikes Peak (the elevation $z = 4{,}300$ m; $\rho/\rho_0 = 0.605$), will the length of your drive be substantially increased? To help you answer these and many other questions about the physics and aerodynamics of golf, you should refer to Bearman and Harvey (1976) and Jorgensen (1994).

Let us go one more step—indeed, a giant leap. What happens if the air density is zero and the gravitational force is greatly reduced? Well, here we are on the moon! We have $\rho_a = 0$ and $g = 1.62\ \text{m/s}^2$. If, again, $U_0 = 60$ m/s and $\alpha = 12°$, you can easily calculate that the range $L = 904\ m$ (instead of 133 m on earth), the maximum height $y_m = 48.0$ m (instead of 11.3 m), and the time of flight $T = 15.4$ s (instead of 3.9 s).

A final point: Are you looking for an interesting topic for a research project? Here is an idea. Everyone has seen those spectacular golf putts on television where the golfer aims the ball at a horizontal angle many degrees to the right or left from the straight line path to the hole. The reason, of course, is that the golfer has studied the topography of the green, the texture of the grass, and other features that will evidently affect the route of the ball. Then the putt is made, the ball moves along a very interesting curved path, and terminates, remarkably, at the cup.

Your assignment for the research project is to devise a suitable mathematical model of this not-easy problem in mechanics. Start with something simple like a sloping plane for the green and uniform friction coefficients. Then go on to more complicated geometries for the green and directional friction coefficients. If you would like some help on this problem, here are two references: Lorensen and Yamrom (1992) and Alessandrini (1995). The first of these is fairly easy reading; the second is somewhat more difficult.

We leave the subject of golf with the suggestion that you might want to read a short article entitled "Mathematics for Golfers" by Stephen Leacock (1956). The author, a professor of political economy at McGill University, wrote a great many funny short stories and essays. This is one of them.

20

Happy Landings in the Snow

In the preceding chapters we examined a number of topics dealing with trajectories of objects moving through a vacuum or through air. We conclude our coverage of the subject with the analysis of another kind of object moving through air: the ski jumper.

As in the case of baseballs and golf balls, important aerodynamic forces are encountered in ski jumping. As the ski jumper accelerates down the in-run ramp and then springs into the air at the end of the take-off table, to begin the free-flight portion of the jump, drag forces and lift forces due to the air are exerted on the jumper and skis. These aerodynamic forces, along with gravity, determine the shape of the flight path or trajectory. The jumper's ability to utilize and control these forces determines the judgment of the style and length of the jump. Much more on all this later.

Modern Ski Jumping

Ski jumping as we know it today evidently had its beginnings in Norway in the 1860s; during the latter part of the nineteenth century the sport spread to Sweden, Finland, Russia, and Germany. Since then, of course, ski jumping has become very popular in a great many countries of the world. In 1924, the Winter Olympic Games were established and held for the first time in Chamonix, France. During the ensuing decades, ever-increasing numbers of nations participated in Olympics ski jumping. At the 1994 games, held in Lillehammer, Norway, fifty-eight skiers from twenty countries competed in ski jumping.

The overall governing body for ski jumping is the International Ski Federation (FIS). This organization determines the rules and regulations for ski jumping and specifies the proper designs and dimensions of ski jump facilities.

A typical FIS "90-meter" ski jump is displayed in figure 20.1. In the figure, a so-called critical point, designated *P*, is shown on the landing slope. The straight line distance from this point to the take-off point *T* is 90 meters; this distance defines the size of the jump. Another point *K*, 20 meters uphill from *P*, identifies the beginning of the landing slope. Most jumpers are expected to land between *K* and *P*. Finally, a point *A* specifies the end of the so-called out-run. At Lillehammer, there are two ski jump facilities: a 90-meter hill and a 120-meter hill.

The starting point of the in-run is located at *E*. This point may be moved uphill or downhill, depending on weather conditions, to assure the proper range of take-off velocities at the take-off point. The gradient of the in-run is 35°; that of the landing slope

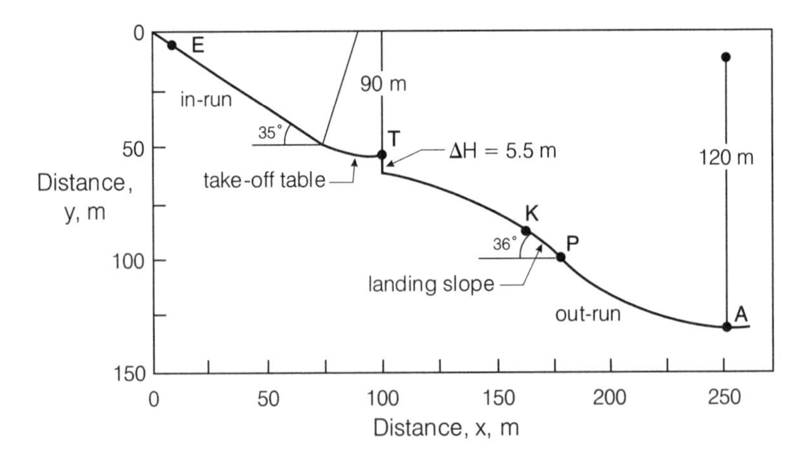

FIG. 20.1

Profile of a 90-meter ski jump facility

is 36°. A vertical curve with 90-meter radius encompasses the take-off table; another vertical curve with 120-meter radius connects points *P* and *A*.

The actual jump consists of five phases: in-run, take-off, free flight, landing, and out-run. Unless a complete debacle occurs on the in-run, judging does not take place until the skier reaches the end of the take-off table. Then the five official judges evaluate the free flight and landing of the jump on the basis of style and length of the jump. Most ski jump experts believe that the best style virtually assures maximum distance.

The In-Run Phase of the Ski Jump

We begin our problem with an examination of the in-run. With regard to figure 20.1, the ski jumper, starting at point *E*, pushes off to a velocity of about three meters per second (which we shall neglect in our analysis) and then immediately goes into what they call the "egg position." Shown schematically in figure 20.2(*a*), this position greatly reduces the aerodynamic drag of the skier during acceleration down the in-run. It was developed by the French a number of years ago from wind tunnel studies. This position also permits the jumper to spring upward and forward at the proper

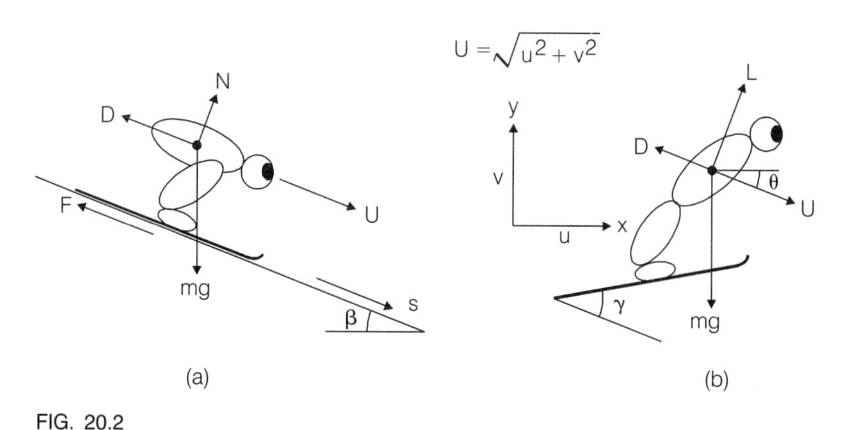

FIG. 20.2

Definition sketches for (*a*) in-run analysis (ski jumper is in the egg position) and (*b*) free-flight analysis (ski jumper is in the vorlage position)

instant during take-off to increase overall velocity; we'll get to that shortly.

First, a word about notation. To simplify the symbols for the various forces, we shall let D = drag force (F_D), L = lift force (F_L), F = friction force (F_F), and so on.

Now for the mathematical analysis of the in-run section of the ski jump. As seen in figure 20.2(a), distance along the in-run ramp is s and the forces acting parallel to this direction are a component of the weight, the friction force of the skis on the snow-covered ramp, and the aerodynamic drag exerted on the skier. Incidentally, for the in-run analysis we neglect aerodynamic lift and the effects of acceleration due to the vertical curve. Also, as in our earlier trajectory analyses, we assume there is no wind.

Accordingly, from the equation of motion we obtain

$$m\frac{dU}{dt} = mg \sin \beta - F - D, \tag{20.1}$$

in which m is the mass, U is the instantaneous velocity, dU/dt is the acceleration, g is the force due to gravity, β is the slope of the in-run, F is the friction force, and D is the aerodynamic drag. The friction force F is given by

$$F = \mu N = \mu mg \cos \beta, \tag{20.2}$$

where μ is the coefficient of friction between the skis and the ramp and N is the component of the weight perpendicular to the ramp.

As in our baseball and golf ball problems, the drag force D is expressed by the equation

$$D = \tfrac{1}{2}\rho_a C_D A^2 U^2, \tag{20.3}$$

in which ρ_a is the density of air, C_D is the drag coefficient, and A is the projected area of the skier. In problems of this type, it is customary to lump C_D and A together and call their product the "drag area," $A_D = C_D A$. Looking ahead, we shall do the same thing for the lift force L with an equivalent "lift area," $A_L =$

$C_L A$. Consequently, we have

$$D = \tfrac{1}{2}\rho_a A_D U^2 \qquad \text{and} \qquad L = \tfrac{1}{2}\rho_a A_L U^2. \tag{20.4}$$

If we substitute these various relationships into equation (20.1) and simplify, we obtain

$$\frac{dU}{dt} = g(\sin\beta - \mu\cos\beta) - \frac{\rho_a A_D}{2m}U^2. \tag{20.5}$$

Letting $a = g(\sin\beta - \mu\cos\beta)$ and $b = \rho_a A_D/2m$, equation (20.5) becomes

$$\frac{dU}{dt} = a - bU^2. \tag{20.6}$$

At this point, we pause to compute a quantity known as the terminal velocity. Suppose that the in-run ramp extends a very long distance down the hill. In fact, we assume the distance is so long that the skier sooner or later reaches a maximum velocity, and so the acceleration becomes zero. Accordingly, setting $dU/dt = 0$ in equation (20.6) and solving for U gives

$$U_* = \sqrt{a/b}, \tag{20.7}$$

in which U_* is the maximum or terminal velocity.

What is a typical numerical value of U_*? To answer this question, it is necessary to calculate the values of the constants a and b, as defined in equation (20.5).

Acceleration of gravity: $g = 9.82\ \text{m/s}^2$.

Slope of the in-run: $\beta = 35°$.

Coefficient of friction: The value of the coefficient of friction, μ, depends on the amount of friction between the skis and the snow. Highly polished skis or special fabrication material for the skis (e.g., teflon) reduce μ to very low values. In our problem, $\mu = 0.05$ is employed.

Density of air: We again utilize the equation we had in the previous chapter to calculate the density of air:

$$\rho_a = \rho_0 e^{-(mg/R_* T)z}, \tag{20.8}$$

in which ρ_0 is the density at sea level ($z = 0$). The other quantities in this equation are $m = 28.96 \times 10^{-3}$, $g = 9.82$, $R_* = 8.314$, and $\rho_0 = 1.205$. During the Winter Games, the temperature was approximately $T = -8°\text{C} = 265°\text{K}$. The elevation of Lillehammer is $z = 240$ m above sea level. Substituting these numbers into equation (20.8) gives the density of air $\rho_a = 1.17 \text{ kg/m}^3$.

Drag area: The drag area for a skier in the egg position on the in-run is given by Ward-Smith and Clements (1982): $A_D = 0.20 \text{ m}^2$.

Mass of the skier: We assume that $m = 75$ kg is the total mass of the skier and skis.

Finally, if we substitute these numerical values into the expressions for a and b, we obtain $a = 5.23$ and $b = 0.00156$. In turn, putting these numbers into equation (20.7) gives $U_* = 57.9$ m/s (208 km/hr). This is the velocity the skier would eventually reach if the in-run ramp were extremely long.

Back to the main problem. In equation (20.6) we let $dU/dt = U(dU/ds)$. So this equation becomes

$$U\frac{dU}{ds} = a - bU^2. \tag{20.9}$$

Separating variables and putting the equation into integral form,

$$\int_0^U \frac{U\,dU}{a - bU^2} = \int_0^s ds. \tag{20.10}$$

The lower limits of the integrals stipulate that the skier's velocity is zero at the start of the in-run, $s = 0$ at point E.

An easy integration of equation (20.10) gives the answer

$$U = U_* \sqrt{1 - e^{-2bs}}\,. \tag{20.11}$$

Note that if $s \to \infty$ then $U = U_*$, as before. Finally, suppose that the length of the in-run is $s = L = 85$ m. Then, with $b = 0.00156$ 1/m and $U_* = 57.9$ m/s, we determine, from equation (20.11), that the ski jumper's velocity on reaching the take-off point is $U_L = 27.9$ m/s (100 km/hr).

PROBLEM. By integrating equation (20.6) directly, demonstrate that the time T of the in-run is given by the expression

$$T = \frac{1}{2bU_*} \log_e\left(\frac{U_* + U_L}{U_* - U_L}\right). \tag{20.12}$$

Substituting numbers, confirm that $T = 5.8$ s in our problem.

The Take-Off Phase of the Ski Jump

The ski jumper has completed the in-run and is moving very fast across the take-off table. At the exact proper instant, the jumper springs upward a distance h meters from the egg position and leans forward to attain the free-flight position.

We can assume that this upward distance h corresponds to a vertical velocity $U_N = \sqrt{2gh}$. Selecting $h = 0.2$ m gives $U_N = 2.0$ m/s. Consequently, the take-off velocity becomes $U_0 = \sqrt{U_L^2 + U_N^2} = 28.0$ m/s. The take-off angle is $\alpha = 10°$.

The timing of the skier's upward spring or leap is extremely critical. A microsecond too soon or too late would be highly detrimental to the style and length of the ensuing free-flight jump.

The Free-Flight Phase of the Jump

Most expert ski jumpers, as they begin the free-flight phase of the jump, go into what they call the "vorlage position" (meaning "forward lean"). As illustrated in figure 20.2(b), the trained skier leans forward, legs straight and arms held at the sides. During free flight, the skier strives to hold a constant angle of attack or incidence, γ, with respect to the trajectory direction. Many excellent jumpers angle their skis into a kind of vee or delta shape in order to obtain additional aerodynamic lift.

The free-flight phase is the most critically judged portion of the jump. It is also the most crucial as far as attainment of the maximum distance is concerned.

Our mathematical analysis of the free-flight trajectory again starts with Newton's equation of motion. Without much trouble we obtain the expressions

$$mu\frac{du}{dx} = -L\sin\theta - D\cos\theta, \tag{20.13}$$

and

$$mv\frac{dv}{dy} = L\cos\theta - D\sin\theta - mg, \tag{20.14}$$

where, as seen in figure 20.2(b), u is the velocity component in the x (horizontal) direction and v is the velocity component in the y (vertical) direction. The instantaneous velocity of the jumper is $U = \sqrt{u^2 + v^2}$ and the angle of the trajectory with respect to the horizontal is θ.

It is convenient to write these equations in the form

$$mu\frac{du}{dx} = -\frac{\rho_a U^2}{2}(A_L\sin\theta - A_D\cos\theta), \tag{20.15}$$

and

$$mv\frac{dv}{dy} = \frac{\rho_a U^2}{2}(A_L\cos\theta - A_D\sin\theta) - mg, \tag{20.16}$$

in which A_L and A_D are the lift area and drag area, respectively. The numerical values of these quantities depend on the angle of incidence γ. Table 20.1 lists values of A_L and A_D for the three incidence angles we shall use in our analysis. These values, obtained from wind tunnel measurements, are given by Ward-Smith and Clements (1982).

Finally, at the mathematical origin $x = 0$, $y = 0$, we know that the take-off velocity is $U = U_0 = 28.0$ m/s and the take-off angle is $\theta = \alpha = 10°$.

It should come as no surprise that it is impossible to obtain exact solutions to equations (20.15) and (20.16). These nonlinear relationships are similar to those we had in the baseball and golf

TABLE 20.1

Numerical values of the lift area A_L and drag area A_D for a ski jumper in the vorlage position

Angle of incidence γ	*Lift area* A_L, m^2	*Drag area* A_D, m^2
8°	0.10	0.40
40°	0.30	1.10
60°	0.35	1.60

Source: Data from Ward-Smith and Clements (1982).

ball trajectory problems we considered in earlier analyses. Consequently, it is necessary to turn to numerical solutions as we did before. For this we utilize the methodology and answers provided by Ward-Smith and Clements (1983), who employed a Runge–Kutta computation procedure to handle the problem.

These investigators obtained numerical solutions for three values of the angle of incidence γ: 8°, 40°, and 60°. The computed trajectories are shown in figure 20.3; the main results of the computations are listed in table 20.2.

Along with the style of the jump, evaluated by the five judges, the length of the jump is all important. So a word about that. For the $\gamma = 8°$ jump, the coordinates of the landing point are $x_L = 101$ m, $y_L = -62$ m. Accordingly, the designated length of the jump is $L = \sqrt{x_L^2 + y_L^2} = 119$ m. In the same way, $L = 94$ m for the $\gamma = 40°$ jump and $L = 75$ m for the $T = 60°$ jump.

The Landing and Out-Run Phases of the Jump

On nearing the completion of free flight, the ski jumper bends at the knees and extends one ski slightly ahead of the other. This so-called "telemark position" provides the jumper with a highly necessary shock absorber on hitting the landing slope. The judges identify the point midway between the two feet as the landing point of the jump; from this, the length of the jump is determined. After landing, the jumper straightens up and continues

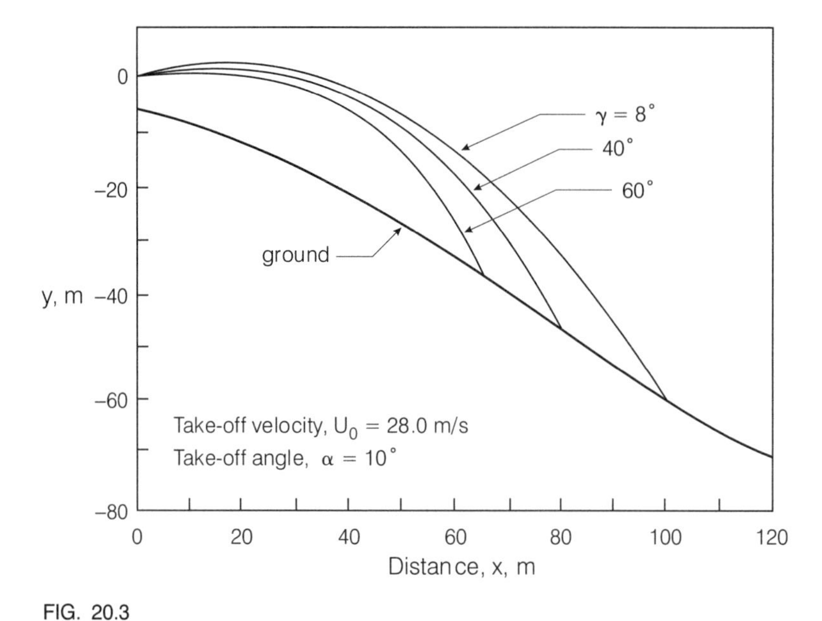

FIG. 20.3

Ski jump trajectories for various values of angle of incidence, γ. (From Ward-Smith and Clements 1983.)

down the landing slope and out-run. After decelerating to a suitable velocity, the skier snowplows or turns to stop.

Some Other Studies of Skiing and Ski Jumping

The subjects of skiing and ski jumping have been mathematically analyzed by de Mestre (1990) and by Townend (1984). An interesting study by Krylov and Remizov (1974) examined ski jumping as a problem in optimum control. They concluded that maximum distance can be attained by minimizing drag area during the initial part of the jump and maximizing lift during the latter part. Comprehensive coverage of many aspects of the physics of skiing is presented by Lind and Sanders (1996).

Various topics on the aerodynamics of skiing are discussed by Raine (1970), by Perlman (1984), and by Epstein (1984). A

TABLE 20.2

Summary of results of ski jump trajectory computations obtained from Runge–Kutta analysis

Parameter	Angle of incidence		
	$\gamma = 8°$	$\gamma = 40°$	$\gamma = 60°$
Lift area $A_L = C_L A$	0.10	0.30	0.35
Drag area $A_D = C_D A$	0.40	0.10	1.60
Designated length of jump L, m	119	94	75
Time in free flight T, s	4.4	4.2	3.4
Designated velocity L/T, m/s	27.0	22.4	22.1
Computed length of trajectory S, m	129	103	83
Average velocity S/T, m/s	29.3	24.5	24.4
Maximum height above ground, m	19.6	17.6	15.7
Minimum free-flight velocity, m/s	24.6	22.0	20.0
Landing velocity, m/s	36.7	27.7	23.7
Landing angle on the landing slope			
with respect to horizontal	−56°	−58°	−60°
with respect to ground	20°	22°	24°

Source: Based on examples given by Ward-Smith and Clements (1983).
Notes: Take-off velocity $U_0 = 28.0$ m/s, take-off angle $\alpha = 10°$, mass $m = 75$ kg, air density $\rho_a = 1.17$ kg/m^3.

delightful little book entitled *The Fantasy of Flow*, published by the Visualization Society of Japan (1993), presents numerous high-speed color photographs of skiers and ski jumpers, as well as spinning and nonspinning golf balls, baseballs, and tennis balls.

21

Water Waves and Falling Dominoes

In this chapter we are going to explore a few topics involving *waves* and *wave motion*. We will not be able to do much more than scratch the surface; the scope of the subject is extremely broad.

Think of how many different kinds of waves there seem to be and how commonplace and crucial many of them are in our everyday lives. Here is a list of various kinds of waves—to which you can probably add some:

light waves	flood waves
sound waves	traffic waves
electromagnetic waves	atmospheric waves
ocean waves	epidemic waves
earthquake waves	population waves

Over a very long period of time—indeed over many centuries—all of these wave phenomena have been studied by a great many mathematicians, scientists, and engineers. As a result, we now have an enormous amount of knowledge and information about waves and wave motion; there is still much more to learn. Highly recommended references dealing with the fundamentals of waves are French (1971a) and Newton (1990).

Water Waves

Since there are so many types of waves, we shall narrow things down quite a bit. Let us consider the topic of "waves on water," that is, the so-called free surface waves that occur on oceans, lakes, and rivers. It is abundantly clear that we still have a very broad and complicated subject.

To begin our consideration of water waves, suppose that wind is blowing across the surface of a body of water and, as a consequence of the wind action, surface waves are created. We could analyze this problem by utilizing the equations for the *conservation of mass* and the *conservation of momentum*. Doing so we would obtain the following equation:

$$\frac{\partial^2 y}{\partial t^2} = C^2 \frac{\partial^2 y}{\partial x^2}, \tag{21.1}$$

in which y is the elevation of the water surface above or below an average value, x is distance, t is time, and C is the velocity of the waves. This equation is known as the *wave equation*; it is one of the most important in all of mathematical physics. It should be mentioned that this equation applies not just to waves on water but indeed to most of the kinds of waves we listed above.

This equation is very important to applied mathematicians and others interested in wave motion. It is an example of a *partial differential equation*. This means that the *dependent* variable—y, in this problem—is a function of more than one *independent* variable. In the present problem, there are two independent variables, x and t. In other words, $y = f(x, t)$. Using equation (21.1) and other information, we obtain the following equation for the velocity C of the water waves:

$$C = \sqrt{\left(\frac{g\lambda}{2\pi} + \frac{2\pi\sigma}{\rho\lambda}\right) \tanh \frac{2\pi H}{\lambda}}, \tag{21.2}$$

where, as seen in figure 21.1, λ is the wavelength (i.e., the distance between successive crests or successive troughs of a wave), H is the water depth, g is the gravitational constant, ρ is

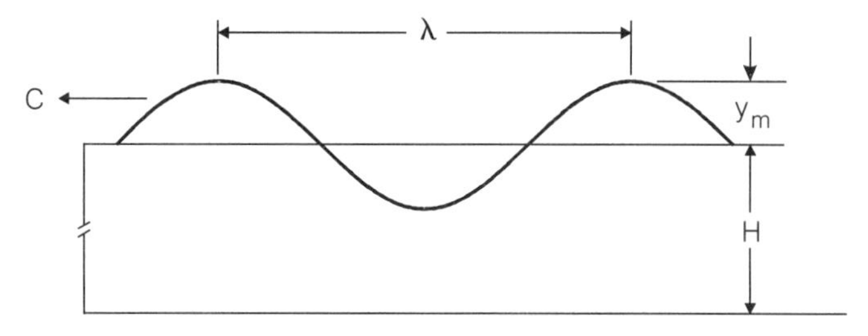

FIG. 21.1

Velocity and length parameters of a free surface wave

the water density, and σ is the surface tension at an air-water interface. The quantity tanh z is the so-called hyperbolic tangent.

Capillary Waves or Ripples

Suppose that our body of water is quite deep, in the sense that the depth H is much larger than the wave length λ, that is, $H/\lambda \gg 1$. In this case, $\tanh(2\pi H/\lambda) = 1$, and so equation (21.2) becomes

$$C = \sqrt{\frac{g\lambda}{2\pi} + \frac{2\pi\sigma}{\rho\lambda}}\,. \tag{21.3}$$

Imagine that a slight breeze is blowing over a water surface. We observe a pattern of ripples of very short wavelength λ. Looking at equation (21.3), we note that if λ is "small," then the first term is negligible in comparison with the second term and so we can drop it. Consequently, the wave velocity in this case is given by the expression

$$C = \sqrt{\frac{2\pi\sigma}{\rho\lambda}}\,. \tag{21.4}$$

This type of water wave is known as the *capillary wave*; clearly, it is caused by the surface tension, σ, of the water. It turns out that the minimum velocity of a capillary wave is $C = 23.2$ cm/s and the corresponding wavelength is $\lambda = 1.73$ cm. Capillary waves are the *ripples* you see when you toss a pebble into a pond of water.

Deep Water Waves

We return to equation (21.2) without the surface tension term. This yields the expression

$$C = \sqrt{\frac{g\lambda}{2\pi} \tanh \frac{2\pi H}{\lambda}}\,. \tag{21.5}$$

Once again, suppose that the water is deep, that is, $H/\lambda \gg 1$. As before, $\tanh(2\pi H/\lambda) = 1$, and hence the wave velocity is

$$C = \sqrt{\frac{g\lambda}{2\pi}}\,. \tag{21.6}$$

This is the wave velocity of *deep water waves*. For example, suppose we observe a wavelength $\lambda = 1{,}000$ ft between the crests of an ocean wave in relatively deep water some distance from the shore. In this case we determine from equation (21.6) that the wave velocity is $C = 72$ ft/s.

Shallow Water Waves

Finally, suppose that the water is no longer deep but, instead, is relatively shallow, that is, $H/\lambda \ll 1$. In this case, $\tanh(2\pi H/\lambda) = 2\pi H/\lambda$, and equation (21.5) becomes

$$C = \sqrt{gH}\,. \tag{21.7}$$

This is the wave velocity of *shallow water waves*. For example, suppose that our deep water wave with wave length $\lambda = 1{,}000$ ft approaches the shore where the depth $H = 50$ ft. In this case, from equation (21.7), the wave velocity is $C = 40$ ft/s.

Deep water waves, shallow water waves, and those in between are the kinds of waves of interest to the oceanographer. However, as you might expect, things are not quite this simple. In the preceding analysis it was necessary to make several assumptions, and one of them was that the ratio of the wave amplitude to the wavelength, y_m/λ, is a small quantity. Consequently, the wave velocity equations we obtained above do not apply to shoaling and breaking waves, nor to things like large-amplitude surges and bores in tidal rivers—which we shall look at shortly.

Tsunamis

A special category of ocean waves is the *tsunami*, a Japanese word that has come to mean a shallow water ocean wave of very long wavelength. These waves are created by submarine and coastal earthquakes and by volcanic eruptions.

Tsunamis generally have a wavelength, λ, of several hundred miles. Indeed, the wavelength is so large that even where the ocean depth is, say, $H = 15{,}000$ ft—the average depth of the Pacific Ocean—the ratio H/λ is sufficiently small to make the tsunami behave as a shallow water wave. For this depth, we find from equation (21.7) that the mean velocity $C = 695$ ft/s = 475 mi/hr.

In the open ocean, the height of a tsunami is never more than a few feet; a ship in the path of a tsunami would not be affected by nor even be aware of its passing. However, as the tsunami approaches a shore, the bottom topography and coastline configuration cause the height of the wave to increase dramatically. When this rapidly moving giant wave hits the shore, it is highly probable there will be catastrophic damage to property and even substantial loss of life.

As we know, the entire rim of the Pacific Ocean is composed of regions of intense earthquake and volcanic activity. So it is not surprising that the Pacific is also the scene of most of the world's tsunamis. You will recall, from chapter 5, that the 1883 explosion of the Krakatoa volcano near Java produced a tsunami over 100 feet high, which brought total destruction to nearby shores of Java and Sumatra and claimed more than 35,000 lives.

Even though most of the major tsunamis of the world have occurred in the Pacific, these catastrophic waves have also done great damage in the Atlantic and Indian Oceans and the Caribbean and Mediterranean Seas.

The devastating effects of a major tsunami can be felt over an extremely large region. For example, in May 1960 a violent earthquake struck Chile, and this was quickly followed by a volcano eruption and a great many landslides. There was great damage and a staggering loss of life in Chile. Moreover, the tsunami created by the earthquake and the associated subsidence of a large undersea fault took a toll on nearly every shoreline of the Pacific Ocean. Coastal cities in New Zealand and even Australia were flooded. The Philippines and Okinawa were hit hard. There was substantial loss of life and great property damage in Japan—nine thousand miles from Chile. West coast cities in the United States, and especially Hilo, Hawaii, experienced enormous damage.

If you would like to learn more about tsunamis, you might read the book by Myles (1985) entitled *The Great Waves*. It describes some of the most destructive tsunamis of the past, including the one created by the Krakatoa volcano explosion in 1883. Two other recommended books are *Earthquakes* by Bolt (1993) and *Waves and Beaches* by Bascom (1980).

Tidal Surges or Bores

Imagine that we have an open channel of rectangular cross section through which water is flowing, as in an irrigation canal. Suppose also that at a certain location along the channel there is a large open gate that spans the channel. For whatever reason, we suddenly close the gate so that the flow of water in the channel is stopped. What happens next is not difficult to visualize. A surge of water, with depth substantially greater than the original depth, begins to move upstream at a certain velocity C. This wave of water is called a surge or a bore. Let us take a look at this interesting kind of water wave.

The surge or bore is a phenomenon that is unlikely to occur very often in an irrigation canal but they do happen frequently in certain rivers of the world. When the topography of the river and the tidal conditions at the river's mouth are "just right," this surge or tidal bore is created and moves in the upstream direction of the river.

Perhaps the most famous bore in the world is the one on the Severn River in England. It has a velocity of about fifteen miles an hour ($C = 22$ ft/s) and a height of four or five feet. The bore on the Seine River in France, between Rouen and Le Havre, has similar characteristics, and the one on the Qiantang River in China can attain a height of fifteen feet. There are also bores on the Petitcodiac River in New Brunswick, Canada, and indeed on the Amazon River in Brazil.

By no means do all tidal rivers have surges or bores. The topographical and tidal characteristics determine whether or not a bore will be created. In connection with the subject of bores and other kinds of water waves, here are two books you might want to examine and study: Le Méhauté (1976) and Tricker (1964). The mathematics involved in the analysis of waves can become extremely complicated; however, these two references are fairly easy reading.

The phenomenon of tidal bores can be analyzed without much difficulty; the problem is defined in figure 21.2. If we again utilize the principles of conservation of mass and conservation of momentum, we obtain the expression

$$C = \sqrt{\frac{gH}{2}\frac{h+H}{h}} - U_1. \tag{21.8}$$

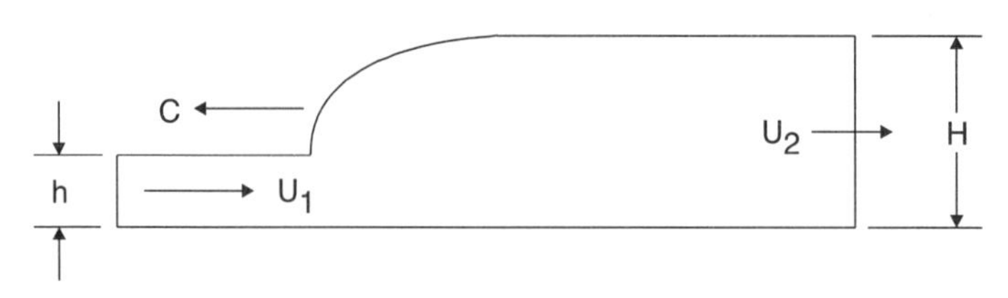

FIG. 21.2

Velocity and length parameters of a surge or tidal bore

Suppose that the bore is moving upstream into a very slowly moving or even quiescent body of water. In this case, we can set $U_1 = 0$. Therefore, equation (21.8) is written in the final form:

$$\frac{C}{\sqrt{gH}} = \sqrt{\frac{1 + (h/H)}{2(h/H)}} \,. \tag{21.9}$$

Suppose, for example, that $H = 15$ ft and $h = 10$ ft. Then this tells us that $C/\sqrt{gH} = 1.118$ and so $C = 24.6$ ft/s $= 16.8$ mi/hr. This is the velocity at which the tidal bore—a large-amplitude water wave—moves upstream.

Falling Dominoes

Now, we have concluded our brief look at tsunamis, bores, and other types of waves on water. We are still interested in "waves," but we now shift our attention to a different and rather strange kind of "wave motion," a long row of falling dominoes.

Everyone has seen those intriguing television scenes in which complicated patterns of standing dominoes are arranged in all sorts of interesting configurations. At time $t = 0$, someone pushes over the first one. Then, for what seems like quite a long time, waves of falling dominoes are moving all over the place—until the last one tumbles over. This is the sort of wave phenomenon we are now going to examine. To keep it simple, we start with a single row of standing dominoes, as lengthy as you like, with a constant spacing between successive dominoes.

We are interested in determining the velocity of the domino wave, that is, the velocity at which the crest of the falling dominoes moves in the direction of the domino row. However, before we begin our analysis of this problem, it is helpful to obtain an answer to the following question.

How Long Does It Take for a Long Pole or a Tall Chimney to Fall Over?

This is a very interesting question. Figure 21.3 serves as a definition sketch. We assume that our pole or chimney has a

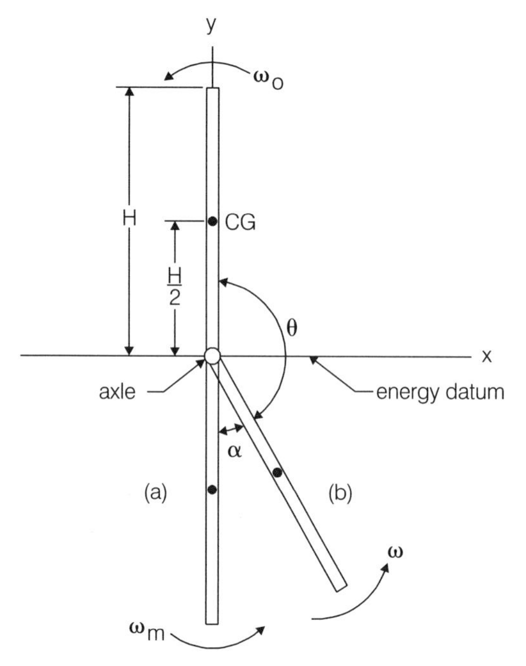

FIG. 21.3

Definition sketch for the falling pole

frictionless axle at its base so that it is free to rotate in a full circle in the (x, y) plane, that is, in the plane of the paper. As the pole rotates, it is assumed that axle friction and air resistance can be neglected.

First, a few definitions. The length of the pole or chimney is H, its diameter is D, its mass is m, and its weight is $W = mg$, where g is the gravitational force ($g = 32.2$ ft/s$^2 = 982$ cm/s^2). A quantity known as the moment of inertia is very important in the study of rotating objects. It is a measure of how mass is distributed in the object. The moment of inertia of our pole about its base is $I = (1/3)mH^2$.

The *potential energy* of an object is given by the expression $E_p = Wy$, where W is its weight and y is the distance of its center of gravity above or below a specified energy datum. The *kinetic energy* of a rotating object is described by the relationship $E_k =$

$(1/2)I\omega^2$, in which I is the moment of inertia of the object and ω is its angular velocity.

We now write down the equation for the conservation of energy between position (a), in which the pole is exactly upside down, its angular velocity is ω_m, and its deflection angle $\alpha = 0$, and position (b), where the pole is now moving upward in the counterclockwise direction with angular velocity ω and deflection angle α. The angle θ is $\theta = 180° - \alpha$.

So the energy conservation equation is

$$W\left(-\frac{H}{2}\right) + \frac{1}{2}I\omega_m^2 = W\left(-\frac{H}{2}\cos\alpha\right) + \frac{1}{2}I\omega^2. \qquad (21.10)$$

The negative signs in the two potential energy terms indicate that the center of gravity is *below* the energy datum. Using $W = mg$ and $I = (1/3)mH^2$, we easily obtain the expression

$$\omega = \sqrt{\omega_m^2 - \frac{3g}{H}(1 - \cos\alpha)}\,. \qquad (21.11)$$

We note that when $\alpha = 0$, $\omega = \omega_m$; this is the *maximum* angular velocity of the pole during its cycle of rotation. From this equation we also obtain

$$\omega_0 = \sqrt{\omega_m^2 - \frac{6g}{H}} \qquad \text{when } \alpha = 180°. \qquad (21.12)$$

This is the *minimum* angular velocity. Finally,

$$\omega_* = \sqrt{\omega_m^2 - \frac{3g}{H}} \qquad \text{when } \alpha = 90°. \qquad (21.13)$$

As we shall see, this is the angular velocity of the pole at the instant it hits the ground, $y = 0$.

Now, by definition, $\omega = d\alpha/dt$, and so equation (21.11) becomes

$$\frac{d\alpha}{\sqrt{1 - (3g/\omega_m^2 H)(1 - \cos\alpha)}} = \omega_m\,dt. \qquad (21.14)$$

Using the trigonometric identity $\sin(\alpha/2) = \sqrt{(1/2)(1 - \cos\alpha)}$, defining $\phi = \alpha/2$, and letting $k^2 = 6g/\omega_m^2 H$, this relationship is written in the final integrated form,

$$\int_0^{\phi} \frac{d\phi}{\sqrt{1 - k^2 \sin^2\phi}} = \frac{1}{2}\omega_m t. \tag{21.15}$$

This equation gives the time, t, for the pole to rotate from $\alpha = 0$ to any angle $\phi = \alpha/2$. In particular, if $\alpha = 180°$ (i.e., π radians), we get

$$\int_0^{\pi/2} \frac{d\phi}{\sqrt{1 - k^2 \sin^2\phi}} = \frac{1}{2}\omega_m t. \tag{21.16}$$

These relationships, equations (21.15) and (21.16), are, respectively, the incomplete and complete elliptic integrals of the first kind. You may remember that the same integrals appeared in chapter 14. We express these equations in the standard form:

$$K(k) = \tfrac{1}{2}\omega_m t_2 \tag{21.17}$$

and

$$F(\phi, k) = \tfrac{1}{2}\omega_m t_2, \tag{21.18}$$

where t_1 is the time required for the pole to rotate from $\alpha = 0$ to $\alpha = 180°$ and t_2 is the time required for it to rotate from $\alpha = 0$ to any angle, α. In particular, for our present problem, we want $\alpha = 90°$.

Finally, after we determine the times t_1 and t_2, the difference, $t_* = t_1 - t_2$, is the time required for the pole to rotate from $\alpha = 90°$ (i.e., $0 = 90°$) to $\alpha = 180°$ (i.e., $\theta = 0°$). Since we are dealing with a frictionless system—the sum of the potential energy and kinetic energy is always constant—t_* is also the time required for the pole to rotate from $\alpha = 180°$ ($\theta = 0°$) to $\alpha = 90°$ ($\theta = 90°$). This may appear to be a roundabout way to get the answer; this approach is necessary because the tables for $K(k)$ and $F(\phi, k)$ have been set up this way.

So how long does it take for the pole to fall over? The answer is given by the equation

$$t_* = \frac{2}{\omega_m}[K(k) - F(45°, k)]. \tag{21.19}$$

There is still one uncertainty. With reference to figure 21.3, let us suppose that initially our pole, sitting on a frictionless axle, is pointing straight upward. It is in so-called neutral equilibrium, and theoretically it will stay in that position forever unless we give it a little push—just enough to create a small angular velocity, ω_0, which we presume we know. We then compute ω_m from equation (21.12) and hence calculate the time of fall, t_*, from equation (21.19).

Time-of-Fall

PROBLEM 1. We have a pole with $H = 25$ ft and $g = 32.2$ ft / s^2. In order to determine ω_m, suppose we assume that the initial angular velocity, ω_0, is a small fraction of ω_m; that is, $\omega_0 = n\omega_m$, where, for example, $n = 0.10$. If we substitute this into equation (21.12) and solve for ω_m, we get

$$\omega_m = \sqrt{\frac{1}{(1 - n^2)} \frac{6g}{H}}. \tag{21.20}$$

With $H = 25$ ft, $g = 32.2$ ft / s^2, and $n = 0.10$, the maximum angular velocity $\mu_m = 2.79$ rad / s. Since $k = \sqrt{6g/H} / \omega_m$, then $k = 0.995$. From a table of complete elliptic integrals of the first kind, it is determined that $K(0.995) = 3.6960$, and so we obtain $t_1 = 2.646$ s.

Incidentally, $t_1 = T/2$, where T is the period of one complete rotation of the pole about the axle.

From a table of incomplete elliptic integrals of the first kind, we also establish the value $F(45°, 0.995) = 0.8000$ and so we find that $t_2 = 0.630$ s. Therefore, $t_* = t_1 - t_2 = 2.016$ s. We conclude that it takes about 2.0 seconds for a 25-foot pole to fall over.

PROBLEM 2. This problem involves a chimney or tall tree: $H = 250$ ft, $g = 32.2$ ft/s^2. If $n = 0.10$, determine that $t_* = 6.4$ s, and if $n = 0.05$, establish that $t_* = 7.9$ s.

PROBLEM 3. If you had to do a lot of these "time of fall" calculations, you might want to have a swifter methodology. Assuming that $n = \omega_0/\omega_m = 0.10$, obtain the following approximate equation for the time of fall, t_*, of a tall pole or slender tree:

$$t_* \text{ (seconds)} = 0.40\sqrt{H \text{ (feet)}}\,. \qquad (21.21)$$

For example, a 300-foot redwood tree falls over in about 7.0 seconds.

What Is the Wave Velocity of a Row of Falling Dominoes?

We are now ready to return to our main problem. We have a long row of standing dominoes on a horizontal surface. The height of a single domino is H, its thickness is D, and its width is B. As seen in figure 21.4, the dominoes have a spacing L between centers. The various angles and angular velocities are the same as those of figure 21.3 relating to the previous analysis.

It is assumed that the domino thickness, D, is small compared to its height, H, but not necessarily small compared to spacing,

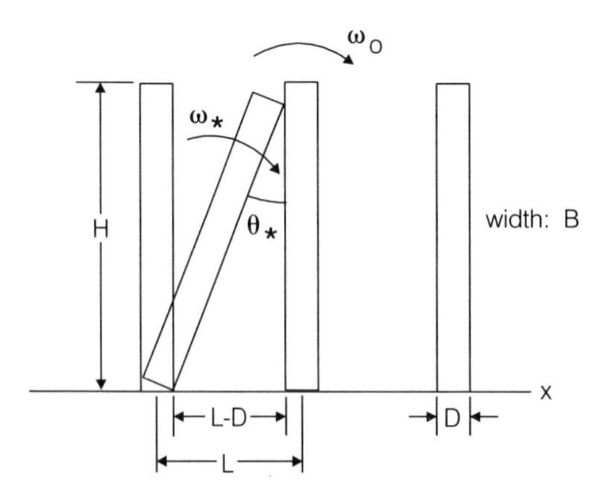

FIG. 21.4

Definition sketch for falling dominoes

L. As before, the moment of inertia of a domino about its base is $I = (1/3)mH^2$. We also assume that only one domino affects the next one in the row. The effects of the preceding three or four already fallen leaning dominoes are neglected. The angular velocity of the falling domino at the instant it hits the next one is ω_*. As seen in figure 21.4 the angle at the instant of impact is θ_*. Clearly, $\sin\theta_* = (L - D)/H$.

The assumption that the energy of a standing domino ($\theta = 0$) is equal to its energy at the instant of impact ($\theta = \theta_*$) gives the expression

$$\omega_* = \sqrt{\omega_m^2 - \frac{3g}{H}(1 + \cos\theta_*)} . \tag{21.22}$$

As before, we also have

$$\omega_0 = \sqrt{\omega_m^2 - \frac{6g}{H}} . \tag{21.23}$$

Now we make our final assumption, and it is very important. We assume that the momentum M in the x direction—the direction of the domino row—is conserved. Since $M = I\omega$, this assumption implies that

$$\omega_* \cos\theta_* = \omega_0. \tag{21.24}$$

Using these expressions, we obtain the following equation:

$$\omega_m = \frac{1}{\sin\theta_*}\sqrt{\frac{3g}{H}\left[2 - (1 + \cos\theta_*)\cos^2\theta_*\right]} . \tag{21.25}$$

As before, we have the elliptic integral relationships that provide the equation

$$t_* = \tfrac{1}{2}\omega_m[K(k) - F(\phi_*, k)], \tag{21.26}$$

in which t_* is the time of fall from $\theta = 0$ to $\theta = \theta_*$, and also $\phi_* = \alpha_*/2 = (180° - \theta_*)/2$. As before also, $k = \sqrt{6g/H}/\omega_m$.

Finally, we define the wave velocity, $C = (L - D)/t_*$. Our computations are completed by calculating the value of the dimensionless parameter $C/\sqrt{gH}$. You will note that this is the same parameter we had in our earlier analysis involving tsunamis, bores, and other kinds of water waves. The *dependent* variable in our mathematical problem is $C/\sqrt{gH}$ and the *independent* variable is the spacing ratio, L/H. The goal of our mathematical problem is the determination of $C/\sqrt{gH} = f(L/H)$.

The results of computations are displayed as the solid curve in figure 21.5. The small open circles and dashed line show the results of experiments carried out by McLachlan et al. (1983).

Some comments. What happens when $\theta_* = 0$, that is, when the gap between successive dominoes is zero? To answer this, we use the series approximations for small values of θ_*: $\sin\theta_* = \theta_*$ and $\cos\theta_* = 1 - \theta_*^2/2$. We substitute these expressions into equation (21.25) and then let $\theta_* = 0$. From this we establish that

$$\omega_m = \sqrt{15g/2H}, \qquad \omega_0 = \sqrt{3g/2H}, \qquad k = \sqrt{4/5},$$

and

$$C/\sqrt{gH} = (1/2)\sqrt{15/2} = 1.369.$$

Clearly, for this limiting *closed-spacing* case, $L/H = D/H$.

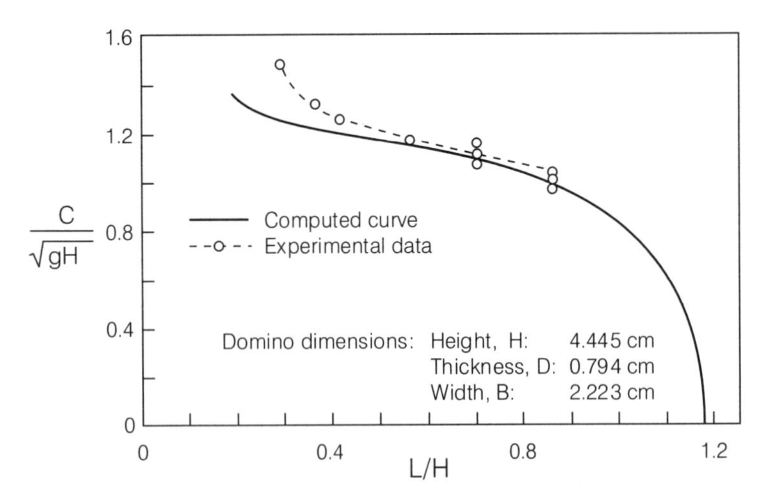

FIG. 21.5

The wave velocity of falling dominoes. Comparison of results of mathematical model with experimental data of McLachlan et al. (1983).

For the limiting *open-spacing* case, $L/H = 1 + D/H$. In this instance, $C/\sqrt{gH} = 0$.

It is seen in figure 21.5 that the results of our simple mathematical model agree fairly well with the experimental data. Even so, you might want to try to devise a better model. In such a model you may want to include not just the effect of a single falling domino, but rather the effects of the few or several already fallen but still leaning smooth dominoes. This kind of analysis, involving a chain of fifteen dominoes, has been carried out by Shaw (1978). By the way, two excellent and not too difficult books dealing with the mechanics of motion of solid bodies are Bullen (1965) and French (1971b).

We conclude our chapter with the following bit of interesting information. In 1988, students of several technical universities in the Netherlands set up and toppled, by a single push, a total of 1,382,101 dominoes. Now suppose that all of these dominoes had been arranged in a single row with a spacing $L = 2.0$ cm between successive dominoes. The total length of the row would be 27.64 kilometers. If the domino height $H = 4.445$ cm, then the spacing ratio $L/H = 0.45$, and so, from figure 21.5, $C/\sqrt{gH} = 1.20$. Hence, $C = 79$ cm/s $= 2.844$ km/hr. So it would take about 9.72 hours for all the dominoes to fall over.

22

Something Shocking about Highway Traffic

What? Shock Waves on Highways?

In this chapter we are going to look at some topics concerning phenomena of wave action on roads and highways. Not many people realize that such things exist. When we hear about wave motion, compression waves, expansion waves, shock waves, and the like, we usually think about ocean waves, sound waves, or even earthquake waves. Yet nearly every time we drive along the highway or expressway we experience "traffic waves" in one form or another. So let's start our chapter with the following little episode of life involving you and your car.

> You are about two hundred yards from a stop light when it changes from green to red. The cars in front of you begin to slow down and stop, and then it is your turn to do so. The density of cars along the particular stretch of road has suddenly increased. Well, this is a nice little compression wave. Successively, the cars in front of you come to a halt, red brake lights flash on, and then yours do and so do the brake lights of the cars behind you. Do you know that you and your fellow drivers just got hit by a red-colored shock wave moving in the opposite direction?

The traffic light changes from red to green and the cars ahead of you begin to move. After what seems a long time, it is your turn to do so. Another wave just passed through you and the other drivers, only this time it was not so dramatic because there are no green accelerator lights on cars like the red brake lights.

In any event, as you begin to get close to the stop light, Murphy's law changes it back to red. With justified chagrin and perhaps a few appropriate words, you come to a fast halt. Again, your red brake lights flash on to provide the starting tracer for another upstream-directed traffic shock wave.

Never mind. You are now at the head of the line. When the green light comes on, you and your car will be the first ones in the next expansion wave.

Determining the Capacities of Highways

During the past several decades, much research has been done to determine the amount of traffic a particular highway can carry. That is, what is the "capacity" of a certain highway? Specifically, this means: How many vehicles can move along a particular stretch of highway per hour and what is the average speed and spacing of the vehicles?

Broadly speaking, this research has fallen into two main categories. The first category has involved the acquisition of engineering data on existing highways. For example, it might be reported that under certain conditions, the capacity of a single lane of a particular highway is 2,400 vehicles per hour, moving at a speed of 30 miles per hour, with an average spacing of 66 feet between vehicles. If you are interested in this kind of technical information on highway capacities, the comprehensive reference book by Hamburger (1982) is a good place to start.

The second category of research dealing with capacities of highways has been theoretical and mathematical in nature. This work has been directed at efforts to develop basic principles of "traffic science" and to devise mathematical models of "traffic flow theories." Examples of endeavors along these lines are the following theoretical and mathematical models:

queuing-line models
car-following models
kinetic theory of gas models
diffusion-dispersion models
fluid flow analogy models

We are not going to consider the first four topics on this list; we shall deal only with the last one: fluid flow analogy models. However, if you want to look into the entire matter more closely, Ashton (1966), Haight (1963), and Prigogine and Herman (1971) are suggested references.

The Fluid Flow Analogy Model of Traffic Flow

Quite a few years ago, Lighthill and Whitham (1955) published a lengthy paper dealing with the theory of highway traffic flow. The basic idea in their approach to the problem is that the flow of traffic along a highway is analogous to the flow of a fluid in an open channel or pipe. This point of view replaces a long column of closely spaced discrete moving vehicles with an equivalent continuous moving stream of liquid (e.g., water) or gas (e.g., air). In other words, Lighthill and Whitham analyzed the phenomenon of traffic flow as though it were a problem in fluid mechanics. This approach allows some, though certainly not all, of the physical and mathematical relationships of hydrodynamics and aerodynamics to be utilized in the traffic flow problem.

From this point of view, our analysis begins with the equation of conservation of mass; a definition sketch is shown in figure 22.1. Consider a short stretch of single-lane highway of length dx, as shown in the figure. The rate at which vehicles cross the left-hand boundary of the stretch (plane 1) is Q vehicles per hour. This quantity, Q, is called the flux or flow rate; in general, it varies with position, x, and time, t. Utilizing the first two terms of a Taylor's series, the rate at which vehicles cross the right-hand boundary of the stretch (plane 2) is $Q + (\partial Q/\partial x)\, dx$.

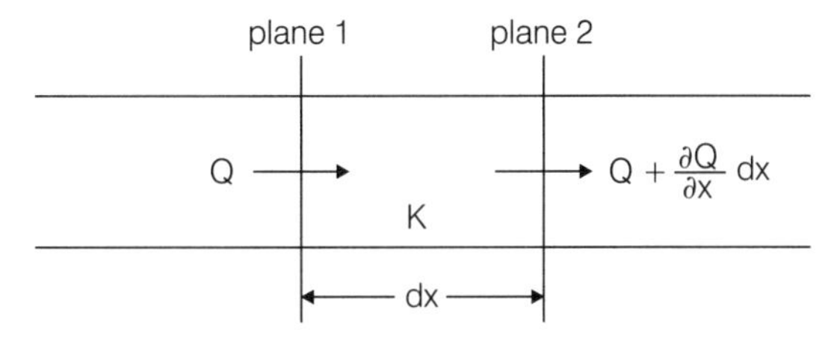

FIG. 22.1

Definition sketch for the conservation of mass; this yields the equation of continuity

Now the density or concentration of vehicles along the highway is K vehicles per mile; in general, this quantity also depends on x and t. At any instant, the number of vehicles within the stretch of highway dx (i.e., between planes 1 and 2) is $K\,dx$, and so the rate at which the number of vehicles is accumulating within the stretch is $\partial(K\,dx)/\partial t$.

From the principle of conservation of mass, we have

$$(Q) - \left(Q + \frac{\partial Q}{\partial x}\,dx\right) = \frac{\partial K}{\partial t}\,dx.$$

The bracketed terms on the left-hand side of this equation are, respectively, the rates at which vehicles are entering and leaving the highway stretch. The right-hand side of the equation is the rate at which vehicles are accumulating within the stretch. Simplifying, this equation becomes

$$\frac{\partial Q}{\partial x} + \frac{\partial K}{\partial t} = 0. \tag{22.1}$$

This equation is called the equation of continuity. The partial derivatives of this equation indicate there are *two independent* variables in the problem, x and t, and that both of the *dependent* variables, Q and K, are functions of x and t, that is, $Q = Q(x, t)$ and $K = K(x, t)$.

At this point, another dependent variable is introduced: the flow velocity, U. This is the average speed of the vehicles; it is defined as follows:

$$U \frac{\text{miles}}{\text{hour}} = \frac{Q \text{ (vehicles/hour)}}{K \text{ (vehicles/mile)}}. \tag{22.2}$$

Again, velocity U depends on position x and time t, that is, $U = U(x, t)$. From this expression we obtain

$$Q = UK, \tag{22.3}$$

which indicates that the flow Q is equal to the product of the velocity U and concentration K.

The Fundamental Diagram of Traffic Flow

Numerous field studies have shown that the velocity U of a stream of traffic on a highway decreases as the concentration of vehicles, K, increases. In our analysis, we shall use the following simple linear relationship between U and K:

$$U = U_* \left(1 - \frac{K}{K_*}\right), \tag{22.4}$$

in which U_* is the maximum velocity (perhaps set by the speed limit or weather conditions) and K_* is called the jam concentration. This equation says that when the concentration $K = 0$, the velocity is a maximum, and when the concentration $K = K_*$ ("bumper to bumper"), the velocity is zero.

Now if we substitute equation (22.4) into equation (22.3), we obtain

$$Q = U_* K \left(1 - \frac{K}{K_*}\right). \tag{22.5}$$

This expression indicates that flow, Q, is zero when $K = 0$ (there are no vehicles on the highway) and also when $K = K_*$ (the traffic is completely jammed and stalled and so the velocity is zero). In our analysis, we assume that within a designated section of highway all vehicles have the same velocity and that no passing is allowed.

Plots of equations (22.4) and (22.5) are shown in figures 22.2(*a*) and 22.2(*b*), respectively. Note that equation (22.5) is a parabola, as shown in the figure. Differentiating this equation with respect to K and setting the result equal to zero shows that the maximum flow, $Q_m = U_* K_* /4$, occurs when the concentration $K_m = K_* /2$. The quantity Q_m is termed the *capacity* of the highway. The relationship between flow Q and concentration K shown in figure 22.2(*b*) is called the *fundamental diagram of traffic flow*.

We pause a moment to look at some numbers. A typical value of the capacity Q_m for single-lane traffic is Q_m = 2,000 vehicles per hour, and a typical jam density is K_* = 160 vehicles per mile. Accordingly, the optimum concentration is K_m = 80 vehicles/mile. Since $Q_m = U_* K_* /4$, the maximum velocity is U_* = 50 miles/hour. On this basis, the traffic is moving at a velocity $U_m = U_* /2$ = 25 miles/hour at a concentration K_m = 80 vehicles/mile to produce the single-lane capacity $Q_m = U_m K_m$ = 2,000 vehicles/hour. Traffic experts indicate that this quantity can usually be multiplied by the number of lanes for multilane highways.

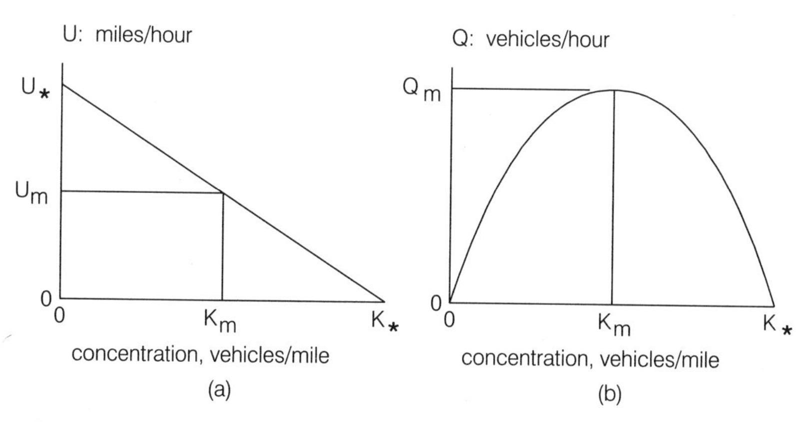

FIG. 22.2

Plots of (*a*) the velocity-concentration curve and (*b*) the flow-concentration curve. This plot is usually called the fundamental diagram of traffic flow.

Kinematic Waves and Characteristic Curves

Back to our mathematical problem. Up to here, we have an equation of continuity, equation (22.1), and what might be called an equation of state, equation (22.5). Rewriting equation (22.1), we get

$$\frac{\partial K}{\partial t} + \frac{dQ}{dK}\frac{\partial K}{\partial x} = 0, \qquad (22.6)$$

or, in alternative form,

$$\frac{\partial K}{\partial t} + C\frac{\partial K}{\partial x} = 0, \qquad (22.7)$$

where $C = dQ/dK$ is the velocity of a kinematic wave moving through the stream of vehicles. Differentiating equation (22.5) gives

$$C = \frac{dQ}{dK} = U_*\left(1 - \frac{2K}{K_*}\right). \qquad (22.8)$$

In addition, we have

$$C = \frac{dQ}{dK} = d(UK) = +K\frac{dU}{dK}. \qquad (22.9)$$

With regard to this equation, we note that dU/dK is a negative quantity [i.e., the slope of the velocity-concentration curve of Fig. 22.2(a) is negative]. Therefore, the wave velocity C is always less than the vehicle velocity U. This means that kinematic waves always move backward through the stream of traffic.

The fundamental diagram of traffic flow is again shown in figure 22.3. We now give geometrical interpretations to some of the preceding quantities.

Consider any point P on the flow-concentration curve. The slope of the radius vector from the origin to point P gives the velocity of the vehicles, $V = Q/K$. Likewise, the slope of the tangent to the curve at point P gives the wave velocity, $C = dQ/dK$. From figure 22.3, we note that if $K < K_m$ (i.e., light

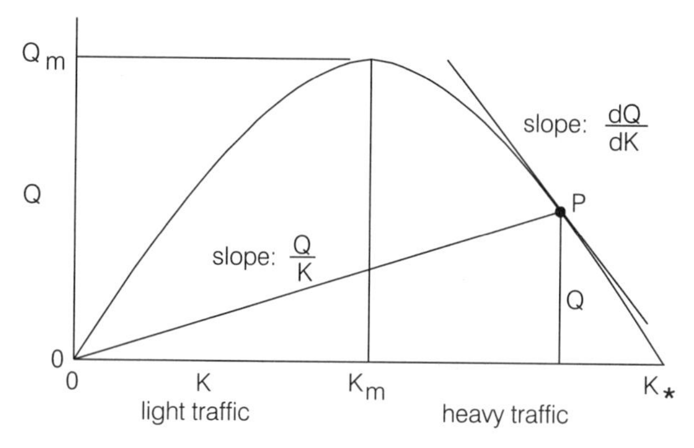

FIG. 22.3

Definition sketch for traffic variables. (*a*) slope of the radius vector, *OP*, gives the vehicle velocity *U*, and (*b*) slope of the tangent at *P* gives the wave velocity *C*.

traffic), this slope is positive and so the wave velocity is positive. Accordingly, the wave moves downstream with respect to the highway (i.e., in the direction of traffic flow). On the other hand, if $K > K_m$ (heavy traffic), the slope and wave velocity are negative and hence the wave moves upstream with respect to the highway (i.e., in the direction opposite to traffic flow). If $K = K_m$, the wave is stationary.

We return to equation (22.7). This is called a *nonlinear first-order partial differential equation.* The general solution to this equation is

$$K = f(x - Ct). \tag{22.10}$$

Thus, the concentration K is a function of the quantity $(x - Ct)$. This quantity represents a so-called kinematic wave moving with velocity C corresponding to a particular value of concentration K. If we plot the position of the wave on a plane with coordinates x and t, we generate a family of straight lines called *characteristics*. The slope of each characteristic is established by the value of the local wave velocity, $C = dQ/dK$. Along each characteristic, the concentration K is constant and equal to its initial

($t = 0$) value. To summarize, the solution to the partial differential equation, equation (22.7), is given by a family of characteristic lines in the (x, t) plane. Each line corresponds to the value of K established by the initial condition of the problem. Detailed presentations of general solutions to these kinds of differential equations are given by Whitham (1974) and by Rhee et al. (1986).

Finally, traffic waves are referred to as *kinematic* waves because they are mathematically generated without bringing any force relationships into the problem. If we were dealing with ocean waves or tidal bores, for example, we would have utilized Newton's second law of motion, $\Sigma F = ma$, in our mathematical analysis. In that case we would be talking about *dynamic* waves. At the present time, unfortunately, there is no law corresponding to Newton's equation for dealing with traffic problems.

Shock Waves in Highway Traffic

Suppose there is a stretch of highway along which the traffic has a relatively high concentration, followed by a stream of traffic in which the concentration is substantially less. This situation may have been caused by an accident, a construction or repair project, a merging side road, or simply a curiosity slowdown. In any event, we examine this traffic situation by turning to the diagram of traffic flow shown in figure 22.4(a).

The more congested stretch of the highway is identified as point 2 in the figure. The slope of the (Q, K) curve at point 2 (line 2) gives the wave velocity C_2. The straight lines shown in the upper part of the (x, t) plot of figure 22.4(b) are parallel to the tangent 2 line. These are the so-called characteristics; each line represents a kinematic wave corresponding to traffic concentration K_2.

The less congested stream of traffic is shown as point 1 in the figure, and the slope of the (Q, K) curve at point 1 (line 1) yields the wave velocity C_1. Again, the kinematic waves shown as the characteristics in the lower part of figure 22.4(b) are parallel to

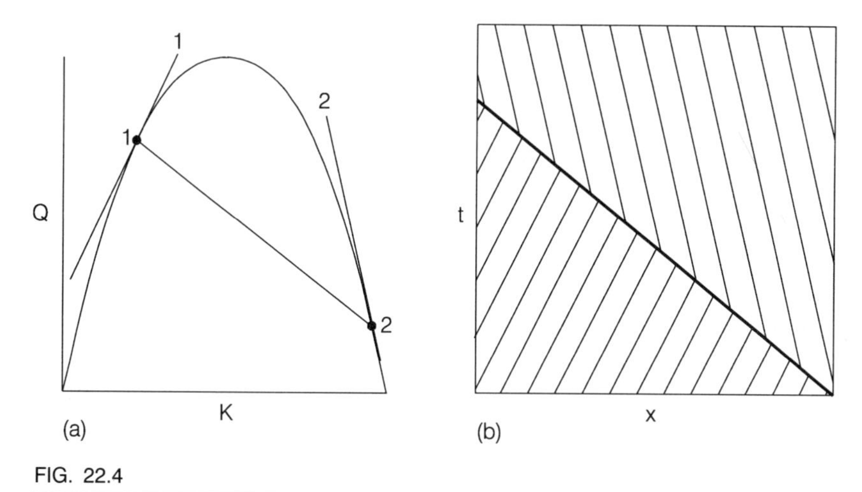

FIG. 22.4

A congested stretch of highway (point 2) followed by a less congested stream of traffic (point 1). (*a*) The fundamental diagram of traffic flow and (*b*) the two families of characteristics and the shock wave.

this slope. When the C_1 and C_2 waves meet, or, in other words, when the two sets of characteristic lines intersect, a *shock wave* is created. The velocity of the shock wave is

$$C_* = \frac{Q_2 - Q_1}{K_2 - K_1}. \qquad (22.11)$$

This velocity is equal to the slope of the chord connecting points 1 and 2 in the figure. In the (x, t) plane of figure 22.4(b), the shock wave is parallel to this connecting chord. Incidentally, as the distance between points 1 and 2 on the (Q, K) curve shrinks to zero, the chord connecting 1 and 2 becomes a tangent to the curve, and the shock wave velocity given by equation (22.11) reduces to $C = dQ/kK$, the velocity of a kinematic wave.

It is shown by Lighthill and Whitham (1955) that the shock waves created by colliding or coalescing kinematic traffic waves have exact analogies in the shock waves confronted in high-speed aerodynamics. Furthermore, the traffic shock wave has an exact counterpart in hydrodynamics in the moving hydraulic jump or so-called tidal bore.

An Example: Brake Lights and Screeching Halts

Anyone who drives a car has the frequent experience of being involved in unexpected slowdowns in highway traffic, crawling along at low velocity for quite a while, and then gradually or suddenly being able to resume a previous high speed. Well, much of this series of events is contained in the scenario we just looked at involving congested and less congested traffic. This time it may be helpful to use some numbers; we use figure 22.5 in our analysis and discussion.

We use the following numerical values in the example:

capacity Q_m = 2,000 vehicles/hour
maximum velocity U_* = 50 miles/hour
jam concentration K_* = 160 vehicles/mile

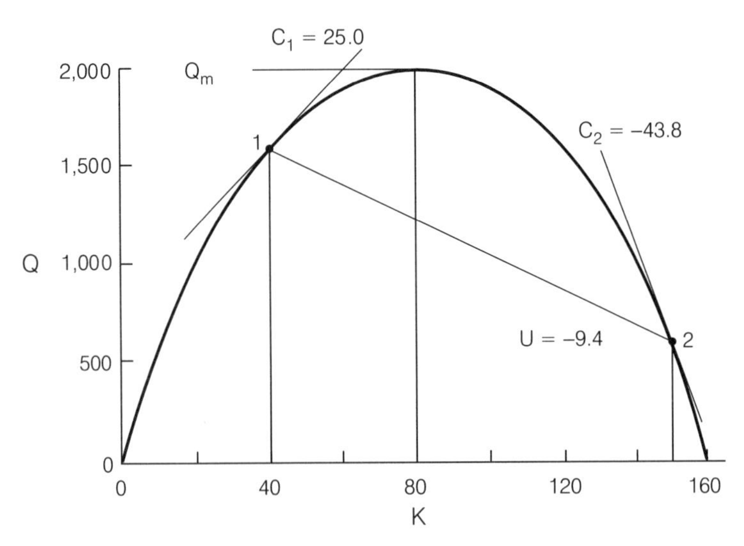

FIG. 22.5

A numerical example. Point 2 ($K_2 = 150$) identifies the congested stretch of the highway. Point 1 ($K_1 = 40$) refers to the less congested stream of approaching traffic.

Summarizing, we have the following relationships:

eq. (22.2), $U = Q/K$
eq. (22.5), $Q = U_* K(1 - K/K_*)$
eq. (22.8), $C = U_*(1 - 2K/K_*)$
eq. (22.11), $C_* = (Q_2 - Q_1)/(K_2 - K_1)$

The following are the results of calculations:

Congested stretch (point 2)
$K_2 = 150$ vehicles/mile
$Q_2 = 469$ vehicles/hour
$U_2 = 3.1$ miles/hour
$C_2 = -43.8$ miles/hour

Approach stream (point 1)
$K_1 = 40$ vehicles/mile
$Q_1 = 1{,}500$ vehicles/hour
$U_1 = 37.5$ miles/hour
$C_1 = 25.0$ miles/hour
$C_* = 9.4$ miles/hour

It is noted that vehicles are moving at a velocity of 37.5 miles per hour before they reach the congested stretch where the traffic is crawling at a velocity of 3.1 miles per hour. The kinematic wave corresponding to the less congested approaching traffic moves *downstream* (i.e., in the direction of the traffic) at a velocity of 25.0 miles/hour. In contrast, the kinematic wave corresponding to the congested stretch moves *upstream* at a velocity of 43.8 miles/hour.

When the two kinematic waves collide, they form a shock wave which moves *upstream* at a velocity of 9.4 miles/hour or about 14 feet/second. As the relatively fast moving traffic approaches the congested stretch, cars decelerate rapidly as brakes are quickly applied; as brakes are applied, brake lights flash on. An observer standing at the edge of the highway sees the brake lights as they successively light up and thus witnesses the passage of a lovely red shock wave as it travels upstream.

Conclusion

The preceding example involving traffic slowdown and the associated generation of a shock wave is illustrative of the kind of analysis it is possible to carry out on traffic flow problems using kinematic wave theory. The example we went through is one of the simpler cases. Needless to say, much more complicated problems have been analyzed. For example, Lighthill and Whitham (1955) have analyzed the following:

progress of a traffic hump
flow through bottlenecks of varying capacity
flow of traffic at junctions
crawl of traffic created by an overcrowded light

If you are interested in learning more about traffic wave theory, good places to start are the books by Beltrami (1987) and Haberman (1977). More advanced treatments are given by Whitham (1974) and by Rhee et al. (1986). By all means, study the original paper by Lighthill and Whitham (1955).

23

How Tall Will I Grow?

The basic question is: How tall will I grow? We begin our answer to this question with a presentation of the results of measurements of the heights H of a large number of American boys and girls, as cited by Tanner (1978). These measurements are given in table 23.1. In addition to the height data, the table also shows height velocities, h; this quantity is the annual increase of height of the children. As the table indicates, both H and h depend on time (i.e., age) t.

The title of the table states that the tabulated quantities correspond to "fiftieth percentile" data. This means, for example, that if the heights of 1,000 American boys of the same age are measured, 500 have heights greater than the tabulated amount and 500 have heights smaller than that amount.

The measured heights and height velocities listed in table 23.1 are displayed in figure 23.1. It is observed that the height at birth ($t = 0$) for both boys and girls is $H_0 = 50$ cm. Boys attain maximum height $H_* = 174$ cm at age $t = 17$ or 18. Girls acquire maximum height $H_* = 162$ cm at age $t = 15$ or 16.

Numerous studies on the subject of human growth have reported that the height of young men continues to increase beyond age eighteen but at a very low rate; ultimate heights are attained at around age twenty-five. Ultimate heights of young women are reached at age eighteen or so.

TABLE 23.1

Measured heights, H, and height velocities, h, of American boys and girls

Age	*American boys*		*American girls*	
t, yr	*H, cm*	*h, cm / yr*	*H, cm*	*h, cm / yr*
0	50		50	
1	75	14.0	75	14.0
2	85	9.2	85	9.2
3	94	7.9	93	7.9
4	101	7.0	100	7.0
5	108	6.5	107	6.5
6	115	6.0	113	6.1
7	121	5.7	120	5.8
8	126	5.5	125	5.5
9	132	5.3	131	5.5
10	136	5.2	137	5.5
11	142	5.0	142	6.5
12	146	5.0	149	8.3
13	152	6.7	156	5.6
14	161	9.5	161	2.4
15	168	5.6	162	0.8
16	172	2.7	162	
17	173	1.0	162	
18	174		162	

Source: Data from Tanner (1978).

Notes: Tabulated quantities correspond to fiftieth percentile data.

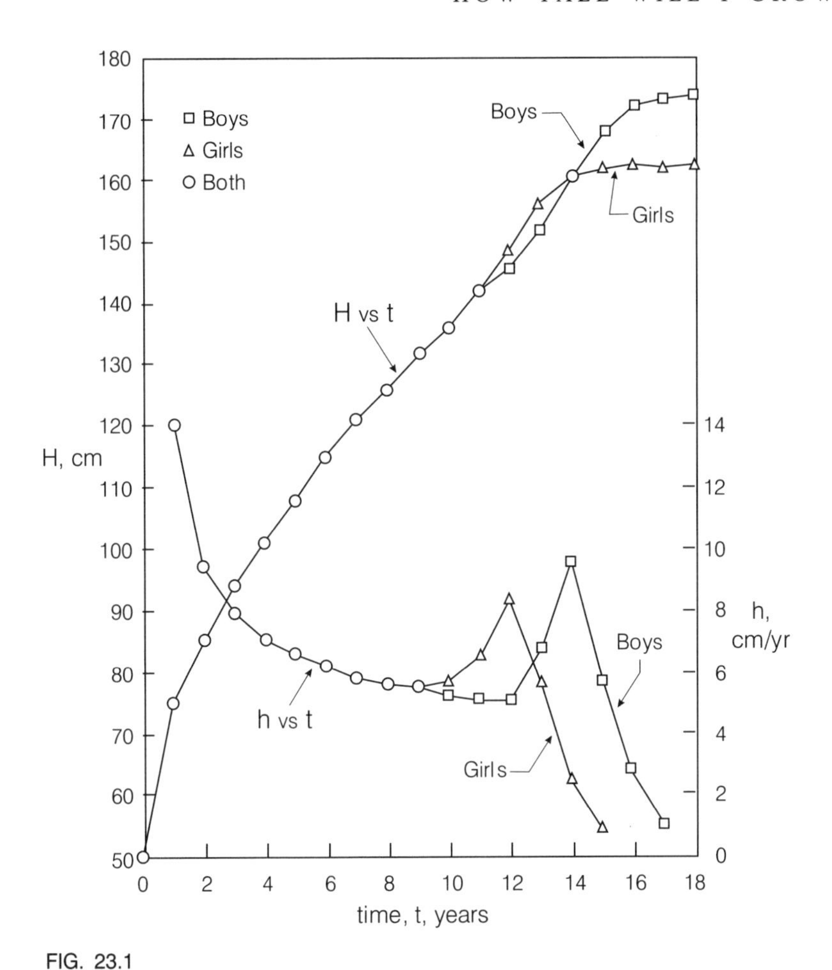

FIG. 23.1

Measured heights, H, and height velocity, h, of American boys and girls. Plotted points correspond to fiftieth percentile data. (From Tanner 1978.)

Remember that in all of this we are considering the 50 percentile group. Later on we shall get to the taller and shorter categories.

As seen in figure 23.1, the annual rate of growth—the so-called height velocity, h—for both boys and girls decreases steadily from $t = 0$ to about $t = 10$. At that time, the growth rate of girls begins to rise and reaches a peak value, $h_* = 8.3$ cm/yr, at

$t = 12$. The growth rate of boys starts to increase at $t = 12$ and attains a maximum value, $h_* = 9.5$ cm/yr, at $t = 14$.

For the benefit of people who want approximate but quick answers to the basic question, reference is made to table 23.2. This table lists, for both boys and girls and at two-year age intervals, the ratio H/H_*, where H is height at age t, and H_* is maximum height, which is attained at age seventeen or eighteen or so.

In the right-hand column of table 23.2, several fractions are shown. These fractions provide approximate answers to the basic question: How tall will I grow? They apply equally to boys and girls, tall and short. The fractions indicate that at age $t = 2$, for example, a child has attained about one-half of his or her final height, at age 4, about three-fifths of final height, and so on.

TABLE 23.2

Values of the ratio H/H_* at two-year age intervals, average values of H/H_*, and approximate fractions of growth, for American boys and girls

	Boys		*Girls*			
Age *t, yr*	*H* *cm*	H/H_* $H_* = 174$ *cm*	*H* *cm*	H/H_* $H_* = 162$ *cm*	*Average* H/H_*	*Approximate fraction of growth*
0	50	0.287	50	0.309		
2	85	0.488	85	0.525	0.507	1/2 (0.500)
4	101	0.580	100	0.617	0.599	3/5 (0.600)
6	115	0.661	113	0.698	0.680	2/3 (0.667)
8	126	0.724	125	0.772	0.748	3/4 (0.750)
10	136	0.782	137	0.846		
12	146	0.839	149	0.920		
14	161	0.925	161	0.994		
16	172	0.989	162	1.000		
18	174	1.000	162	1.000		

An Analysis of the Problem

Take another look at figure 23.1, especially the plot of height H versus time t. We want to try to match or fit this observed correlation between height and time to some kind of mathematical equation. If we are successful in this exercise of "curve fitting," then we can use the mathematical equation to compute various quantities related to growth, including height velocity, acceleration, maximum values, and inflection points.

In the past, many investigators have worked on the problem of obtaining mathematical descriptions of human growth, especially of height and weight. Numerous relationships have been employed as growth functions, including the exponential equation, the logistic equation, the Gompertz equation, and various combinations of these. Descriptions of these efforts are given in technical books by Tanner (1978) and especially Bogin (1988). Incidentally, a quite nontechnical reference on the subject of human growth is the book by Tanner and Taylor (1981).

Our analysis, like those of most previous investigators, begins with the equation

$$\frac{dH}{dt} = a(t)(H_* - H), \tag{23.1}$$

in which H is the height of a person, t is time (i.e., age), and H_* is the maximum or final height. The quantity $a(t)$ is called the growth coefficient; it may be a constant or it may change with time.

This relationship is the differential equation for so-called modified exponential growth. It says that the rate of increase of height, dH/dt, is proportional to the difference between the final height H_* and present height H. Its solution is

$$\log_e \frac{H_* - H_0}{H_* - H} = \int_0^t a(t)\, dt, \tag{23.2}$$

where H_0 is the height at $t = 0$.

Growth Equations for Boys

From the list of height measurements for boys tabulated above, we easily compute the quantity appearing on the left-hand side of equation (23.2); for convenience let $\lambda = \log_e[(H_* - H_0)/(H_* - H)]$. Next we construct a plot of λ versus time t. This is seen in figure 23.2.

What we would like to do is to obtain a single equation, of the simplest possible form, that passes through the entire range of data points shown in the figure. This is not an easy assignment but it could be accomplished if we wanted to take the time.

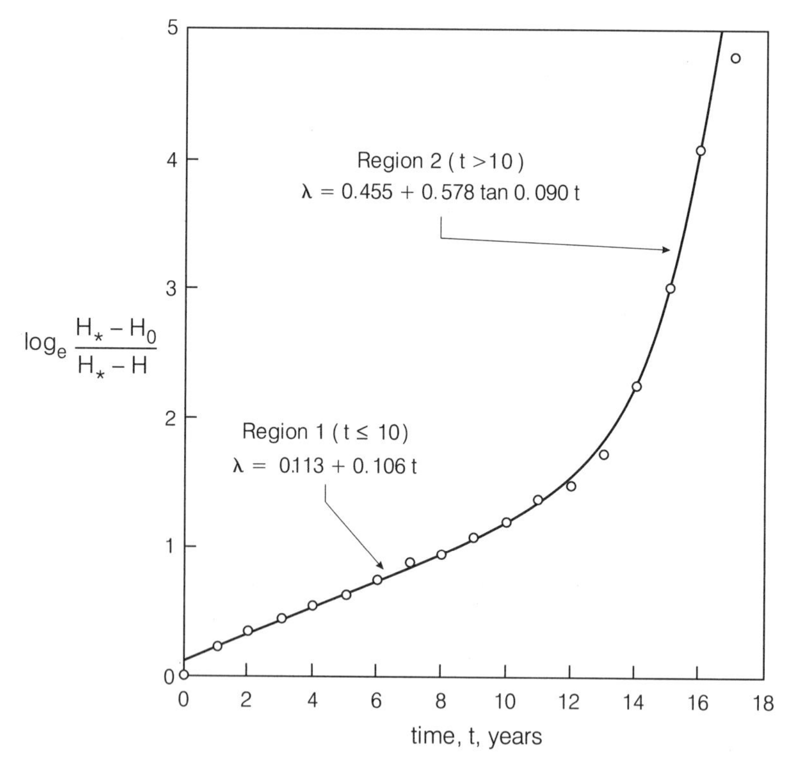

FIG. 23.2

Plot of the height parameter, λ for American boys, fiftieth percentile data. $\lambda = \log_e[(H_* - H_0)/(H_* - H)]$.

Maybe you would like to take on this task as a research topic or possibly a term paper in your science or mathematics course.

Instead, what we shall do is to fit *two* equations to the data points. The first will be for region 1 ($t \leq 10$) and the second for region 2 ($t > 10$).

Region 1

From figure 23.2, it appears that there is linearity between λ and t, that is,

$$\lambda = a_0 + a_1 t, \tag{23.3}$$

where a_0 and a_1 are constants. To determine the values of these constants, it is necessary to carry out a least-squares analysis of the (t, λ) data from $t = 0$ to $t = 10$. This analysis produces a remarkably good linear relationship between λ and t with correlation coefficient $r = 0.9995$, $a_0 = 0.113$, and $a_1 = 0.106$. So we have the result

$$\lambda = 0.113 + 0.106t; \qquad t \leq 10. \tag{23.4}$$

This straight line relationship is shown in the figure.

Region 2

Observe in figure 23.2 that the $\lambda(t)$ curve begins to bend upward at $t = 10$. For curve fitting purposes we could probably match a quadratic equation $(b_0 + b_1 t + b_2 t^2)$ or a cubic equation $(b_0 + b_1 t + b_2 t^2 + b_3 t^3)$ to the data; you might want to look into this. However, from our book on trigonometry we note that the curve of tangent x versus x looks remarkably similar to the pattern of our data for $t > 10$. So, as the framework for our observed data, we try the mathematical expression

$$\lambda = b_0 + b_1 \tan ct, \tag{23.5}$$

in which b_0, b_1, and c are constants. As before, we try to fit the mathematical curve to the data points. This time we obtain the following results: $b_0 = 0.455$, $b_1 = 0.578$, $c = 0.090$. Accordingly, equation (23.5) becomes

$$\lambda = 0.455 + 0.578 \tan 0.090t; \qquad t > 10. \tag{23.6}$$

This equation is also shown in figure 23.2.

Growth Equations for Girls

An identical analysis is carried out with the observed measurements for girls shown in table 23.1. In this case the dividing line between regions 1 and 2 is $t = 8$. Our results are shown in table 23.3.

Mathematical Equations for Height and Height Velocity

Returning to equation (23.2), we write

$$\lambda = \log_e \frac{H_* - H_0}{H_* - H} = \int_0^t a(t)\, dt. \tag{23.7}$$

Although we no longer need to know the precise mathematical form of the growth coefficient $a(t)$, it is not hard to determine it.

TABLE 23.3

Summary of the growth equation coefficients

Region	*Boys* $H_* = 174$ *cm*, $H_0 = 50$ *cm*	*Girls* $H_* = 162$ *cm*, $H_0 = 50$ *cm*
Region 1	$t \leq 10$	$t \leq 8$
$\lambda = a_0 + a_1 t$		
a_0	0.113	0.122
a_1	0.106	0.119
Region 2	$t > 10$	$t > 8$
$\lambda = b_0 + b_1 \tan ct$		
b_0	0.455	0.381
b_1	0.578	0.711
c	0.090	0.098

Note: $\lambda = \log_e[(H_* - H_0)/(H_* - H)]$.

For region 1 we have $\lambda = a_0 + a_1 t$. Substituting this relationship into equation (23.7) and differentiating with respect to time, we obtain the result $a(t) = a_1$. For region 2, $\lambda = b_0 + b_1 \tan ct$. Again, we easily obtain $a(t) = b_1 c \sec^2 ct$, where sec = secant = 1/cosine. This relatively complicated form for the growth coefficient simply reflects our arbitrary selection of the tangent function in equation (23.5).

From equation (23.7), we obtain the following equation:

$$H = H_* - (H_* - H_0)e^{-\lambda}, \tag{23.8}$$

where, as seen in table 23.3, $\lambda = a_0 + a_1 t$ for region 1 and $\lambda = b_0 + b_1 \tan ct$ for region 2. We continue our analysis with an examination of the two regions separately.

Region 1

The height equation in this region is

$$H = H_* - (H_* - H_0)e^{-(a_0 + a_1 t)}. \tag{23.9}$$

Differentiating this expression gives the height velocity,

$$h = \frac{dH}{dt} = a_1(H_* - H_0)e^{-(a_0 + a_1 t)}. \tag{23.10}$$

Recall from differential calculus that the first derivative, dH/dt, expresses the *slope* of the $H(t)$ curve. We note from equation (23.10) that the slope is always positive, that is, the height H increases as time t increases.

Differentiating equation (23.10) provides the height acceleration,

$$\frac{dh}{dt} = \frac{d^2H}{dt^2} = -a_1^2(H_* - H_0)e^{-(a_0 + a_1 t)}. \tag{23.11}$$

Again, recall that the second derivative, d^2H/dt^2, represents the *curvature* of the $H(t)$ curve. We see from equation (23.11) that the curvature is always negative. This means that the $H(t)$ curve is concave downward in region 1. Again, this is in agreement with equation (23.1).

Region 2

In this region the height equation is

$$H = H_* - (H_* - H_0)e^{-(b_0 + b_1 \tan ct)}, \tag{23.12}$$

and so the equation for height velocity is

$$h = \frac{dH}{dt} = b_1 c(H_* - H_0)e^{-(b_0 + b_1 \tan ct)} \sin^2 ct. \tag{23.13}$$

Again, the slope of the $H(t)$ curve is always positive.

The derivative of equation (23.13) provides an equation for the curvature of the $H(t)$ curve in region 2. It is quite a complicated equation, and we have little need for it. The equation does provide, however, an expression for determination of the inflection point. Setting the derivative of equation (23.13) equal to zero yields the equation

$$\tan^2 ct_c - \frac{1}{b_1} \tan ct_c + 1 = 0, \tag{23.14}$$

where t_c is the inflection point time. The solution to this quadratic equation is

$$\tan ct_c = \frac{1}{b_1}\left[1 + \sqrt{1 - b_1^2}\right]. \tag{23.15}$$

From table 23.3, for boys, $b_1 = 0.578$ and $c = 0.090$. Substituting these numbers in equation (23.15) yields $t_c = 14$. For girls, $b_1 = 0.711$ and $c = 0.098$, which gives $t_c = 12$. Both of these inflection point times agree with the plots of figure 23.1. By definition, the inflection point identifies the time of zero height acceleration or, in other words, the time of maximum height velocity.

Computed Curves for Height and Height Velocity

Now we put it all together. For region 1 ($t \leq 10$ for boys; $t \leq 8$ for girls), equation (23.9) is used to compute the height H, and equation (23.10) for the height velocity h, using the numerical constants listed in table 23.3. The results are shown in figure 23.3.

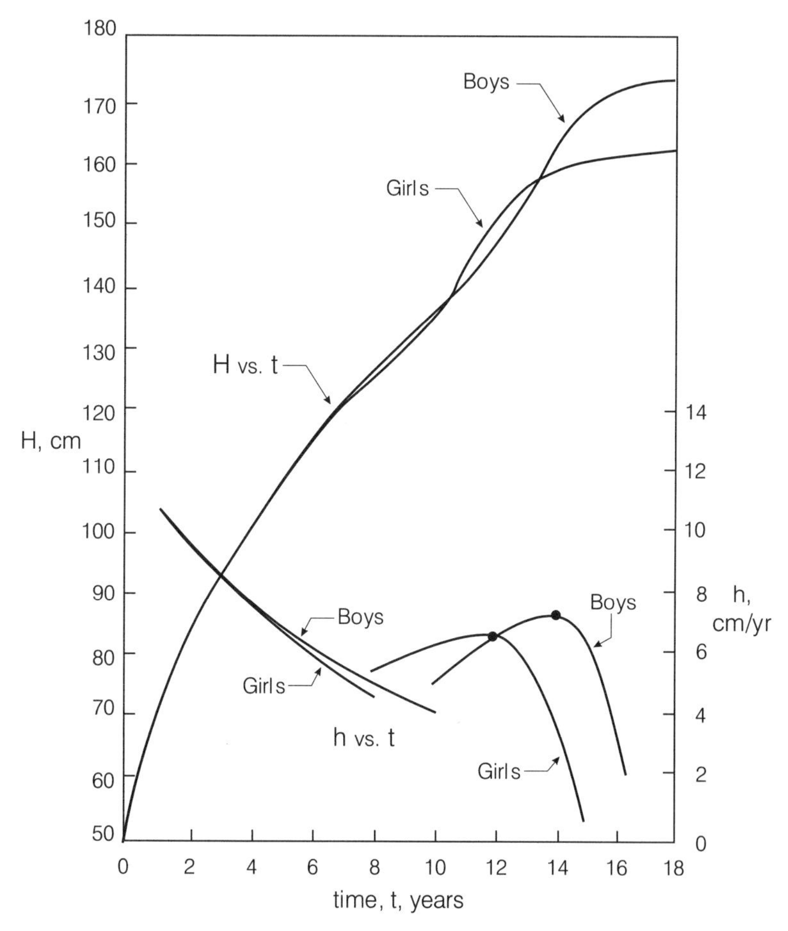

FIG. 23.3

Computed heights, H, and height velocities, h, of American boys and girls. Indicated curves correspond to the fiftieth percentile data listed in table 23.1, plotted, to the same scale, in figure 23.1.

Likewise, for region 2 ($t > 10$ for boys; $t > 8$ for girls), equations (23.12) and (23.13) are employed to calculate H and h, respectively. These curves connect to the region 1 curves at $t = 10$ and $t = 8$. The computed height-time curves of the figure agree well with the tabulated data.

On the other hand, the calculated height velocity–time curves do not entirely match the observed data. The computed times of maximum height velocity, t_c, are satisfactory. In region 1, the calculated height velocities agree rather closely with the observed values. However, in region 2 the computed height velocities are somewhat less than the observed values. The maximum height velocities are considerably less than observed maximum velocities, and they are much less "peaked." Furthermore, there are breaks ("discontinuities") in the height velocity curves at the boundaries between regions 1 and 2. Otherwise, our mathematical model looks pretty good.

Distribution of Heights of Young People

Recall that the observed height measurements seen in table 23.1 and analyzed in subsequent sections refer to the fiftieth percentile category or group of boys and girls. What about the groups that are taller or shorter than this average category? To answer these kinds of questions we need to carry out a statistical analysis.

Suppose we have a group of 1,000 boys all 18 years of age. Obviously, there is a distribution of heights in this group—some boys are quite tall, some are quite short, and others are of various heights in between.

The first thing we do in our statistical analysis of the boys is to prepare a so-called histogram. As shown in figure 23.4, the histogram is simply a bar chart that shows the distribution of the heights. For example, in the $k = 7$ "class interval" of the histogram, it was determined that there are 228 boys with heights between 172 cm and 176 cm.

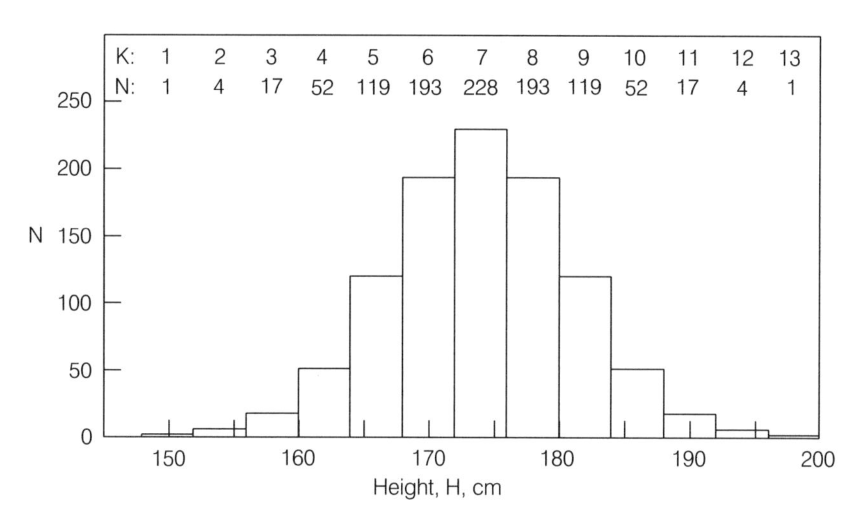

FIG. 23.4

Histogram of the heights of American boys. Age $t = 18$ yr; number in group, $N_0 = 1{,}000$; class interval width, $\Delta H = 4.0$ cm.

Next, we compute the average or mean height, $\overline{H}$, of the 1,000 boys. The formula to calculate this quantity is

$$\overline{H} = \frac{1}{N_0} \sum H_k N_k, \tag{23.16}$$

where N_0 is the total number in the group ($N_0 = 1{,}000$), H_k is the height corresponding to the midpoint of the class interval, and N_k is the number of boys with heights in that interval. For example, for the $k = 7$ class interval, $H = 174$ cm and $N = 228$.

We also want to compute the standard deviation; this index is a measure of how much the distribution of heights is spread out. The equation for calculating this quantity is

$$\sigma^2 = \frac{1}{N_0} \sum \left(H_k - \overline{H}\right)^2 N_k, \tag{23.17}$$

in which σ^2 is the variance; recall that the standard deviation σ is the square root of the variance.

It is interesting that the histogram of equation (23.4) can be described by the normal probability function. It has the definition

$$u = \frac{1}{\sigma\sqrt{2\pi}} \exp\left[-\frac{1}{2}\left(\frac{H - \overline{H}}{\sigma}\right)^2\right], \tag{23.18}$$

where $u = N/N_0$.

If the data presented in the histogram are utilized in equations (23.16) and (23.17), it is not difficult to compute the numerical values of $\overline{H}$ and σ. These values are listed in table 23.4.

Substituting the numbers given in the table into equation (23.18) gives the two curves shown in figure 23.5(a). These "bell-shaped" curves are called the density or frequency distributions of the normal probability function. One of the mathematical properties of this density distribution is that inflection points occur at distances $\pm\sigma$ measured from the mean value $\overline{H}$. Another property is that 68.3% of the total number in the group ($N_0 = 1{,}000$) have heights within $\pm\sigma$ from the mean, 95.5% within $\pm 2\sigma$, and 99.7% within $\pm 3\sigma$.

Finally, we want to compute the total area under the histogram or density distribution curve. In mathematical terms this means that we integrate equation (23.18). The following result is obtained:

$$U = \frac{1}{\sqrt{2\pi}} \int_{-\infty}^{(H-\overline{H})/\sigma} e^{-z^2/2}\, dz, \tag{23.19}$$

where z is a variable of integration. The functional relationship is

TABLE 23.4

Values of mean heights $\overline{H}$ and standard deviations σ at age $t = 18$

Parameter	*Boys*	*Girls*
$\overline{H}$, cm	174	162
σ, cm	7.0	6.0

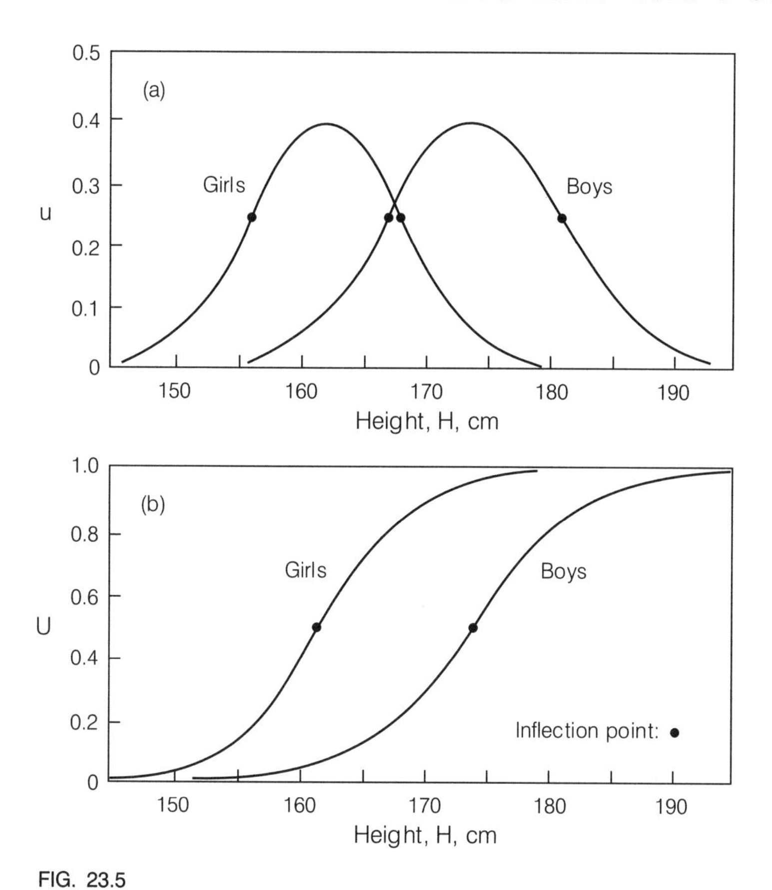

FIG. 23.5

Distribution of heights of American boys and girls. Age $t = 18$. (*a*) Density distribution and (*b*) cumulative distribution.

between U and the upper limit of the integral, $(H - \overline{H})/\sigma$. Plots are shown in figure 23.5(*b*). These "S-shaped" curves are called the cumulative distributions of the normal probability function. The quantity U, defined by equation (23.19), in fact represents the decimal fraction or percentile of the entire group of eighteen-year-old boys (or girls) whose heights are equal to or less than H.

From figure 23.5(*b*) or from tables of the normal probability function, we determine values of H corresponding to specified values of U. These are seen in table 23.5 in the first two columns

TABLE 23.5

Heights, H, and ratios, m, to mean height of eighteen-year old boys and girls, corresponding to percentile U

Boys			Girls			Average
U %	H cm	$m = \frac{H(U\%)}{H(50\%)}$	U %	H cm	$m = \frac{H(U\%)}{H(50\%)}$	$m = \frac{H(U\%)}{H(50\%)}$
97	187	1.075	97	173	1.070	1.07
90	183	1.052	90	170	1.049	1.05
75	179	1.029	75	166	1.025	1.03
50	174	1.000	50	162	1.000	1.00
25	169	0.971	25	158	0.975	0.97
10	165	0.948	10	154	0.951	0.95
3	161	0.925	3	151	0.932	0.93

under each of the headings, "Boys" and "Girls." For example, 90% of eighteen-year-old boys have heights equal to or less than 183 cm; 25% of eighteen-year-old girls have heights equal to or less than 158 cm.

The two columns (one for boys, one for girls) under the heading $m = H(U\%)/H(50\%)$ are simply the decimal fractions of H divided by the mean heights: 174 cm for boys and 162 cm for girls. Since the m values for boys are about the same as those for girls, it is sufficiently accurate to use the average values of m. These are shown in the right-hand column of the table. This is what we are after.

Recall that our entire analysis of height measurements is based on the fiftieth percentile groups of boys and girls. We now use the m values listed in table 23.5 to make corrections to the heights we computed and plotted in figure 23.3. We simply multiply the height values, for all values of age t, by the proper m factors.

Here are a couple of examples. Suppose you have reason to believe that your four-year-old son will eventually be considerably taller than most boys his age. For example, suppose he is in the ninety-seventh percentile group. In this case, from table 23.5, $m = 1.07$. Accordingly, from figure 23.3, at age 10 he is likely to be $136 \times 1.07 = 146$ cm and at age 18 he will reach his final height of 187 cm.

Alternatively, you think that your six-year-old daughter may be somewhat shorter than other girls. Suppose she is in the twenty-fifth percentile group with $m = 0.97$. Again utilizing figure 23.3, we see that at age 12 she will be $150 \times 0.97 = 146$ cm tall. When she reaches age 16 or so, she will attain her final height of 158 cm.

Some Final Problems: Basketball Players and Racehorse Jockeys

PROBLEM 1. A quite tall young woman plans to be a basketball player. Her height is three standard deviations more than the mean height of other young women her age. What is her height and into what percentile does she fall?

Answer. $H = \bar{H} + 3\sigma = 162 + 3(6.0) = 180$ cm (5 ft 11 in). Also, $(H - \bar{H}) / \sigma = 3$. This is the integral upper limit we see in equation (23.19). From tables of the normal probability function, we obtain $U = 0.9986 = 99.86$ percentile. This corresponds to one person in a group of about 715.

PROBLEM 2. A relatively short young man intends to become a racehorse jockey. His height is two standard deviations less than the mean height of other young men his age. What is his height and into what percentile does he fall?

Answer. $H = \bar{H} - 2\sigma = 174 - 2(7.0) = 160$ cm (5 ft 3 in). Further, $(H - \bar{H}) / \sigma = -2$. From the tables we obtain $U = 0.0227 = 2.27$ percentile. This is equivalent to about 23 people per 1,000 or approximately one person in 44.

We conclude that not too many people can become professional basketball players or racehorse jockeys.

24

How Fast Can Runners Run?

It Depends on How Far They Run

Humans have been running and racing for as long as there have been humans. An early example: Prehistoric cavemen, without doubt, displayed remarkable running abilities as they endeavored to widen the gap between themselves and hotly pursuing saber-toothed tigers.

Nearly everyone has heard of the race called the marathon, the length of which is 26 miles and 385 yards. This is the distance a messenger allegedly ran, in the year 490 B.C., from the plains of Marathon to the city of Athens to convey the news of the Greek victory over the Persians. And so on throughout history.

Now even though running and racing have been around for many thousands of years, only during the past hundred years or so has it been possible to measure accurately the time periods and time intervals associated with these activities. Accordingly, it seems accurate and appropriate to suggest that the beginning of modern running and racing essentially coincides with the start of the Olympic Games in Athens in 1896. We should really say "re-start" because the original Olympic Games began in 776 B.C. when Coroibos won the foot race.

Starting with the early years of the twentieth century, a great many studies have been conducted on the physiology and biomechanics of running. Furthermore, and not surprisingly, numerous mathematical models have been formulated over the years.

Without question, the most significant of the early mathematical analyses of running was that of the British physiologist, Archibald Hill. Indeed, for his extensive research on the behavior, action, and fatigue of muscles, Hill was awarded the Nobel prize in physiology in 1922.

Fifty years later, the noted American applied mathematician, Joseph Keller, analyzed the phenomenon of running as a problem in *optimal control theory*. His work, coupled with that of Hill, forms the basis of the so-called Hill–Keller mathematical model of running. We shall examine and utilize this model later in the chapter.

Running Records and Endurance Equations

The first step in our analysis to determine how fast runners can run is to tabulate the current world records of all the official races between 100 meters and the marathon (42,195 meters). There are races shorter than these and races much longer. At the one extreme, there are time-recorded races of 50 yards and 50 meters. At the other extreme, races as long as 100 kilometers, 500 kilometers, and even 1,000 kilometers have been timed. However, we restrict our attention to races between $L = 100$ m and $L = 42{,}195$ m.

The list of current records for men's races is given in table 24.1 and that for women's races in table 24.2.

Next, we make a graph of race distance L versus the time T taken to run this distance. However, instead of simply plotting L versus T, we plot $\log_{10} L$ versus $\log_{10} T$. By plotting the logarithms of these two quantities, we reduce the size of our graph by bringing the data points much closer together.

Using the information listed in the above tables, the two plots shown in figure 24.1 are obtained. It is apparent that there are excellent correlations for the data for both the men's and women's races.

It is reasonable to assume that these correlations can be described by straight lines of the general form $y = k_0 + k_1 x$, or,

TABLE 24.1

Records for men's races

Nominal distance, m	*Distance L, m*	*Nominal time, hr:min:s*	*Time T, s*	*Velocity* $U = L/T$, *m/s*
100	100	9.84	9.84	10.16
200	200	19.32	19.32	10.35
400	400	43.29	43.29	9.24
800	800	1:41.73	101.73	7.86
1,000	1,000	2:12.18	132.18	7.57
1,500	1,500	3:28.86	208.86	7.18
1 mile	1,609	3:44.39	224.39	7.17
2,000	2,000	4:50.81	290.81	6.88
3,000	3,000	7:28.96	448.96	6.68
5,000	5,000	12:56.96	776.96	6.44
10,000	10,000	26:52.23	1,612.23	6.20
20,000	20,000	56:55.6	3,415.6	5.86
25,000	25,000	1:13:55.8	4,435.8	5.64
30,000	30,000	1:29:18.8	5,538.8	5.60
marathon	42,195	2:06:50	7,610	5.54

Source: Data from Wright (1996).

in particular,

$$\log_{10} L = \log_{10} a + c \log_{10} T, \tag{24.1}$$

in which a and c are constants. With some algebra we put this equation into the form

$$L = aT^c, \tag{24.2}$$

where, for men, $a = 12.53$, $c = 0.904$; for women, these constants have the values $a = 12.05$, $c = 0.895$. You can easily modify this expression if you want an equation for T in terms of L.

TABLE 24.2

Records for women's races

Nominal distance, m	*Distance L, m*	*Nominal time, hr:min:s*	*Time T, s*	*Velocity $U = L/T$, m/s*
100	100	10.49	10.49	9.53
200	200	21.34	21.34	9.37
400	400	47.60	47.60	8.40
800	800	1:53.28	113.28	7.06
1,000	1,000	2:30.67	150.67	6.64
1,500	1,500	3:50.46	230.46	6.51
1 mile	1,609	4:15.61	255.61	6.30
2,000	2,000	5:28.69	328.69	6.08
3,000	3,000	8:06.11	486.11	6.17
5,000	5,000	14:37.33	877.33	5.70
10,000	10,000	29:31.78	1,771.78	5.64
marathon	42,195	2:21:06	8,466	4.98

Source: Data from Wright (1996).

Finally, recalling that the average velocity $U = L/T$, equation (24.2) is written in the form

$$U = KL^n, \tag{24.3}$$

in which, for men, $K = 16.38$, $n = 0.106$; for women, these constants are $K = 16.12$, $n = 0.117$. The mathematical relationship defined by equation (24.3), which is sometimes called the "endurance equation," is shown in the plots of figure 24.2 along with the observed data points.

We compute, from equation (24.3), that in a men's race of $L = 100$ m, for example, the average velocity is $U = 10.05$ m/s. This computed value agrees closely with that seen in table 24.1.

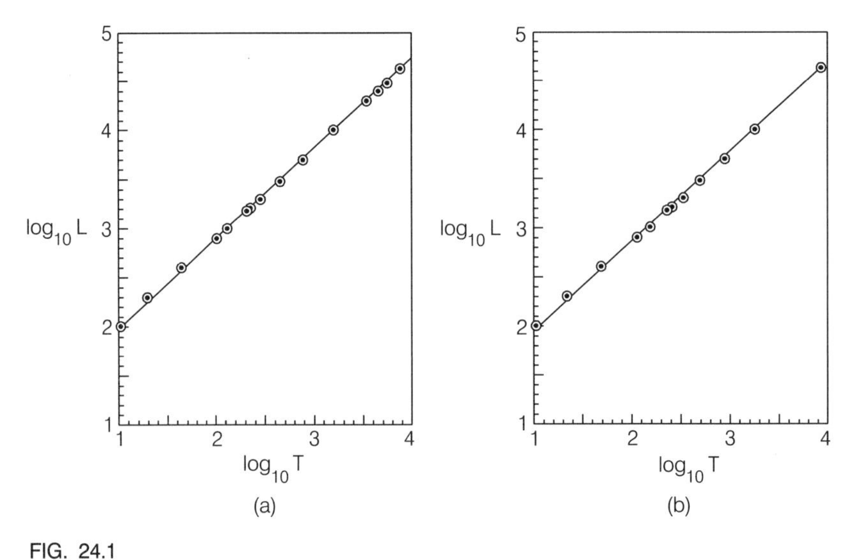

FIG. 24.1

Records for running races. Plots of distance, L, versus time, T, for (*a*) men's races and (*b*) women's races. Common logorithms of both L and T are plotted. Recall that $\log_{10} 10 = 1$, $\log_{10} 100 = 2$, and so on.

Even so, bear in mind that this equation is simply the "best fit" relationship of the observed data. It is not intended to provide exact values of U.

The Hill–Keller Mathematical Model of Running

What we have done up to here is to list the world records of men and women for quite a few races, construct graphical plots of the running data, and prepare empirical equations to describe these data analytically. Now we begin the analysis.

Our mathematical analysis is a simplified version of what has come to be known as the Hill–Keller model of running. It consists of two parts. Part I is the analysis of short dashes or *sprints*, in which the runner exerts maximum propulsive effort and energy expenditure to attain maximum acceleration and velocity

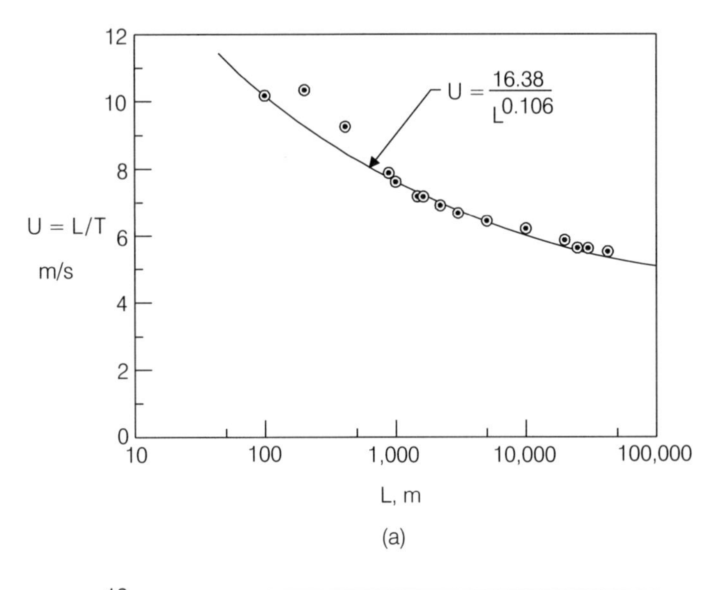

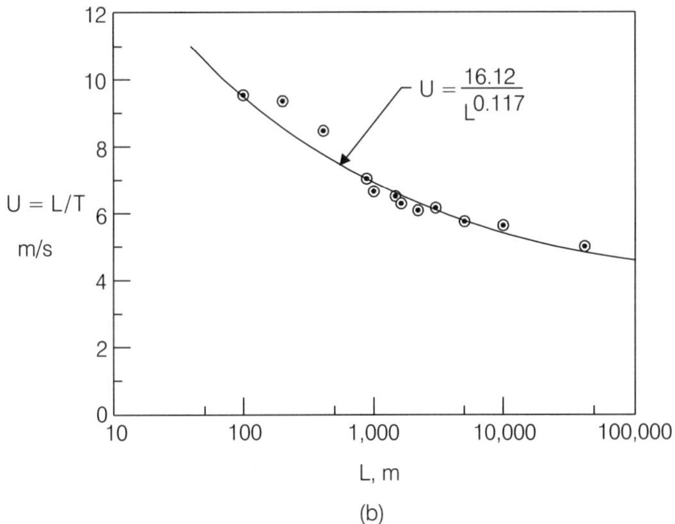

FIG. 24.2

Plots of endurance equations for (*a*) men's races and (*b*) women's races

over a short period of time and distance. Part II is the analysis of *distance runs*, in which the runner, following a brief period of propulsive effort and acceleration, maintains a constant velocity and constant energy expenditure over a prolonged period of time and distance.

Mathematical Analysis of Sprint Races

We begin our analysis with Newton's equation of motion $\Sigma F = ma$, in which ΣF is the summation of all the forces acting on a body of mass, m, to produce an acceleration a. In our analysis, the "body" is the runner whose mass is m.

A propulsive force per unit mass, F, is provided by the leg action of the runner. In addition, there is a restrictive force per unit mass, F_r. This force consists of all the internal factors (e.g., continual acceleration and deceleration of the limbs) and external factors (e.g., track friction) that resist or oppose the running motion. Also, recall that the acceleration a is defined by the relationship $a = du/dt$, in which u is the instantaneous velocity of the runner. So Newton's equation becomes

$$\frac{du}{dt} = F - F_r. \tag{24.4}$$

Based on his extensive studies of the physiology and biomechanics of running, Hill (1927) assumed that the resistive force F_r is directly proportional to the runner's velocity. Virtually every study on running carried out since Hill's early work has made the same assumption. We shall do likewise. Accordingly, we utilize the relationship $F_r = (1/\tau)u$, in which τ is a constant. By the way, most researchers believe that the aerodynamic drag exerted on a runner moving through otherwise still air (i.e., no wind) is only about 3%–6% of the total resistive force. So in our analysis we neglect aerodynamic resistance. Consequently, Newton's equation takes on the form

$$\frac{du}{dt} = F - \frac{1}{\tau}u. \tag{24.5}$$

In sprints, the runner exerts maximum propulsive effort during the entire race. Let's call this maximum propulsive force F_*. Now we have the relationship

$$\frac{du}{dt} = F_* - \frac{1}{\tau}u. \tag{24.6}$$

Since the runner starts with zero velocity, the initial condition is $u(0) = 0$.

The mathematical relationship of equation (24.6) is called the differential equation for modified exponential growth. It says that the acceleration du/dt is a maximum just after $t = 0$ (when $u = 0$), and is continuously reduced thereafter until it becomes zero when $u = u_{max} = F_*\tau$. The solution to equation (24.6) is

$$u = u_{max}(1 - e^{-t/\tau}). \tag{24.7}$$

In addition, $u = dx/dt$, where x is the distance from $x = 0$, the starting point of the race. So we integrate equation (24.7) to get

$$x = F_*\tau^2\left(\frac{t}{\tau} - 1 + e^{-t/\tau}\right). \tag{24.8}$$

At the end of the race, $x = L$ and $t = T$, and equation (24.8) becomes

$$L = F_*\tau^2\left(\frac{T}{\tau} - 1 + e^{-T/\tau}\right). \tag{24.9}$$

It is time for an example. At the World Championship races held in Rome in 1987, Carl Lewis ran the $L = 100$ m race in $T = 9.86$ s. During his race, split time measurements were made of his performance. That is, time intervals were recorded of his position at $x = 0$, $x = 10$ m, $x = 20$ m, and so on to $x = 100$ m. These measurements, provided by Pritchard (1993) and plotted in figure 24.3, show the distance he had run in time t.

The plot of figure 24.3 is mathematically described by equation (24.8). It turns out that the constant τ is sufficiently small to make the exponential term negligible for the larger values of

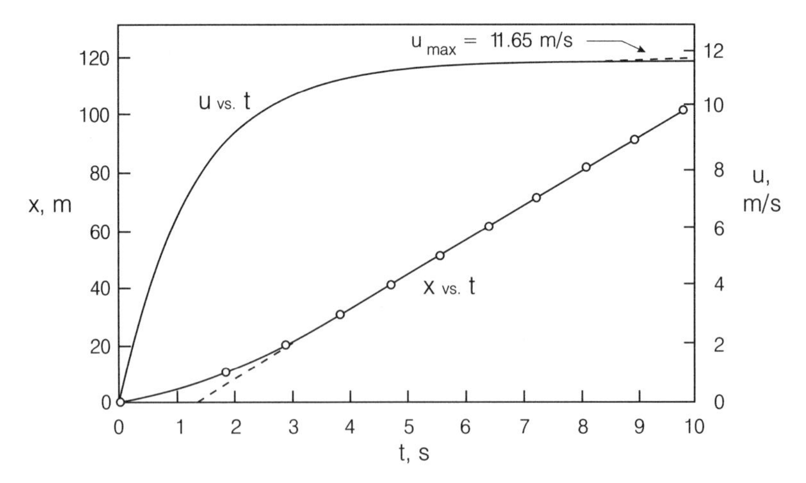

FIG. 24.3

Plot of split time measurements, x vs. t, of $L = 100$ m sprint of Carl Lewis (Rome, 1987). Time $T = 9.86$ s; velocity $U = L/T = 10.14$ m/s. Plot of computed velocity, u vs. t, is also shown. (From Pritchard 1993.)

time, t. Therefore, for $t > 4$ s, equation (24.8) simplifies to the form

$$x = F_* \tau (t - \tau). \tag{24.10}$$

Using this relationship, we easily determine from figure 24.3 that $F_* = 9.19$ newton/kg or m/s^2 and $\tau = 1.27$ s. The maximum velocity is $u_{max} = F_* \tau = 11.65$ m/s. Substituting these numerical values in equation (24.7) gives the computed velocity curve $u(t)$ shown in figure 24.3. Again neglecting the exponential term, equation (24.9) becomes

$$L = F_* \tau (T - \tau). \tag{24.11}$$

Consequently, the average velocity, $U = L/T$, is

$$U = F_* \tau \left(1 - \frac{\tau}{T}\right). \tag{24.12}$$

For Carl Lewis's 100-meter spring, $T = 9.86$ s and so $U = 10.14$ m/s.

Before we leave sprints, here is a good question. At the instant Carl Lewis crossed the finish line of the 100-meter race, what was his kinetic energy? We assume his mass m was 75 kilograms. The kinetic energy of a body of mass m, moving at velocity u, is given by the equation $e = (1/2)mu^2$. Accordingly, the answer to the question is $e = (1/2)(75)(11.65)^2 = 5{,}090$ joules. As you might want to confirm, this is equal to the kinetic energy of a 10-inch diameter steel ball falling from a height of 25 feet.

Mathematical Analysis of Distance Races

In the analysis of races longer than sprints (i.e., races of distances greater than about 200 meters), the Hill–Keller model begins with the energy relationship

$$\frac{dE}{dt} = \sigma - Fu, \tag{24.13}$$

in which $E(t)$ is the energy equivalent per unit mass of the available oxygen in the muscles of the runner, σ is the rate of energy supplied by oxygen due to breathing and circulation in excess of that supplied when the runner is dormant, and Fu is the rate of energy utilized in running. Recall that F is the propulsive force per unit mass of the runner and u is the runner's velocity.

From equation (24.5), we have the following expression for F:

$$F = \frac{du}{dt} + \frac{1}{\tau}u. \tag{24.14}$$

Substituting this relationship into equation (24.13) and integrating with respect to time t yields

$$E(t) = E_0 + \sigma t - \frac{1}{2}v^2(t) - \frac{1}{\tau}\int_0^t v^2(t)\,dt, \tag{24.15}$$

where E_0 is the initial amount of energy in the available oxygen in the muscles.

In his analysis of the problem, Keller (1973, 1974) uses a modified equation (24.15) to indicate that the optimal running velocity for sprint races is identical to equation (24.7). He also shows that this solution is valid for all races of less than a critical distance L_c, which he calculates to be approximately 300 meters.

For races longer than L_c, Keller assumes that the runner's energy is entirely utilized by the end of the race. With this assumption, Keller solves equation (24.15) by methods utilized in the *calculus of variations*. His final answer is quite complicated. However, essentially his solution says that the optimal strategy for the distance runner is to accelerate initially for one or two seconds and then to run at constant velocity for the rest of the race.

A simplified form of Keller's complete solution is given by the expression

$$U = \sqrt{\frac{E_0 \tau}{T} + \sigma \tau}\,, \tag{24.16}$$

in which $U = L/T$ is the average velocity of the runner, E_0 is the runner's initial energy, σ is the energy supply rate, and τ, of course, is the resistive force constant defined by equation (24.5). The approximate solution given by equation (24.16) is the same as the answer originally obtained by Hill many years ago.

We shall not attempt to examine, nor even to introduce, the many complex problems of physiology and biomechanics involved in running. Péronnet and Thibault (1989) give useful information on the subject of the physiology of running and Williams (1985) is a suggested reference on biomechanics. From another point of view, Ward-Smith (1985) analyzes running on the basis of the first law of thermodynamics. Pritchard and Pritchard (1994) present a clear and concise survey article on mathematical models of running. Finally, an easy to read book about running and racing is Brancazio (1984).

It may be interesting and helpful, nevertheless, to introduce a couple of features related to the physiology and biomechanics of running. The first feature is shown in figure 24.4, which presents a plot of running velocity U versus *stride length* λ and *stride frequency* ω. Clearly, $U = \lambda\omega$. For example, if the stride length of a runner is $\lambda = 2.0$ m and the stride frequency is $\omega = 3.0$ per s, then the runner's velocity is 6.0 m/s.

A second feature involving physiology and biomechanics is concerned with heat generation and heat dissipation in distance races. Now virtually all of the energy produced by the runner is dissipated in the form of heat. For a runner of mass m moving at velocity U, the rate of heat generation is $h_g = c_1 m U^2$, where c_1 is a constant. The mass of the runner, m, is proportional to the cube of his or her height H. So $h_g = c_1(c_2 H^3)U^2 = c_3 H^3 U^2$. On the other hand, the rate of heat dissipation is proportional to the surface area of the runner's body, which, in turn, is proportional to H^2. Accordingly, $h_d = c_4 H^2$.

Equating the rate of heat generation to the rate of heat dissipation gives $H^3U^2 = cH^2$, in which c is an overall constant. This result yields $U^2 = c/H$, that is, the velocity is inversely proportional to the height. In other words, in the distance races, such as the marathon, smaller runners have the advantage.

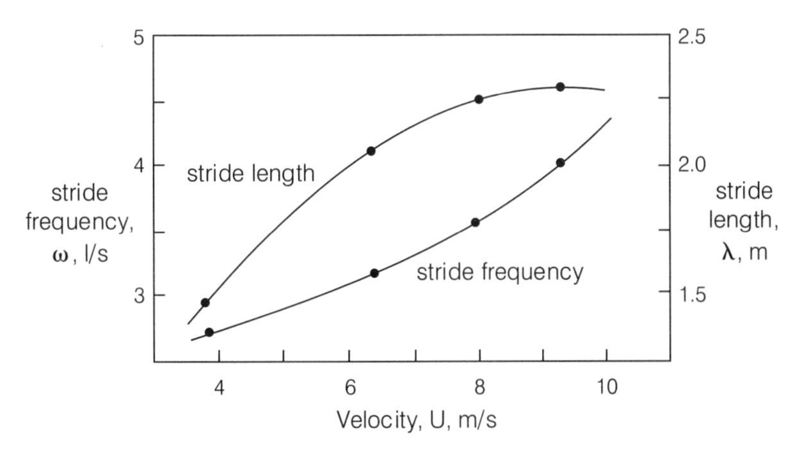

FIG. 24.4

Typical values of stride length, λ, and stride frequency, ω, for various values of running velocity, U. (From Williams 1985.)

The Running Parameters and the Running Curve

Utilizing the information on running records listed in previous tables, the numerical values of the various running parameters are determined. These values are shown in table 24.3.

Just as we did in preparing figure 24.2, we plot the average velocity U versus the distance L for the record races of men and women. The two plots of the data are displayed in figure 24.5. Shown also in the figure are solid curves that present the results of our mathematical model. These curves are called "running curves."

Starting at the left end of the distance coordinate L in figure 24.5, here is what our model tell us:

1. For the short races—the sprints—equation (24.12) gives the expression relating average velocity U to the race time T. That is, repeating equation (24.12) here, we have

$$U = F_* \tau \left(1 - \frac{\tau}{T}\right). \tag{24.12}$$

To obtain the coordinates employed in figure 24.5, that is, $U = f(L)$, we simply use $U = L/T$. This relationship is accurate for race distance L equal to about 200 meters or so.

2. It is seen in figure 24.5 that the average velocity U for both men's and women's races reaches a maximum when L is about 150

TABLE 24.3

Numerical values of the running parameters

Parameter	*Symbol*	*Units*	*Men*	*Women*
Propulsive force	F_*	m/s^2	14.36	13.50
Time constant	τ	s	0.739	0.723
Maximum velocity	$F_* \tau$	m/s	10.61	9.76
Initial energy	E_0	joule/kg	2,790	2,735
Energy rate	σ	joule/kg s	55.6	44.0
Fatigue constant	γ	1/s	6.15×10^{-5}	3.45×10^{-5}

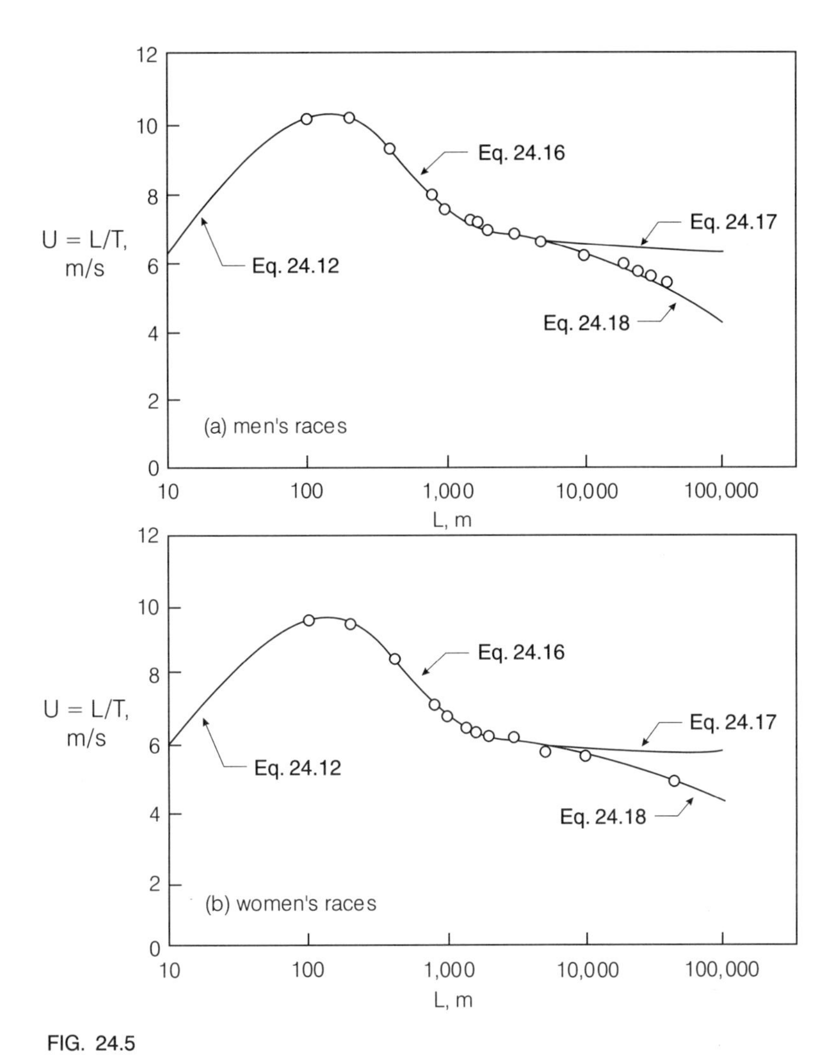

FIG. 24.5

Plots of the running curve for (*a*) men's races and (*b*) women's races. The maximum average velocity, *U*, occurs when the race distance is about 150 meters.

meters. For races shorter than this distance, the runner devotes a sizeable fraction of the race time accelerating from zero velocity at the start. Consequently, the runner's average velocity is lowered. For races longer than 150 meters, the runner must restrict his or her velocity to conserve energy.

3. For races longer than $L = 400$ m, equation (24.16) gives a major result of our mathematical model. For easy reference, it is repeated here:

$$U = \sqrt{\frac{E_0 \tau}{T} + \sigma \tau}\,.$$

This expression, along with $L = UT$, is the equation of the solid line in figure 24.5 for values of L larger than about 400 meters.

4. It is noted that for large values of time T, equation (24.16) reduces to the form

$$U_c = \sqrt{\sigma \tau}\,, \qquad (24.17)$$

which indicates that the running velocity is constant regardless of distance L. This critical velocity, U_c, is established by a balance between energy expenditure due to running and energy replenishment by oxygen intake and consumption. Distance runners refer to this balance as the "aerobic limit." Using the numbers given in table 24.3, we get $U_c = 6.41$ m/s for men and $U_c = 5.64$ m/s for women. From the plot of figure 24.5, we note that this critical velocity can be maintained for distances up to approximately $L = 10{,}000$ m.

5. The power expended by a runner is easily calculated. If the runner is moving at constant velocity U, then $du/dt = 0$, $u = U$, and equation (24.5) shows that the propulsive force per unit mass is $F = U/\tau$. If the runner's mass is m then the total propulsive force is $F_T = mU/\tau$. Since power = force × velocity, the power is $P = mU^2/\tau$. Substituting numerical values gives $P = (60)(6.41)^2/0.739 = 3{,}335$ newton m/s or joules/s or watts = 3.34 kilowatts. Using the appropriate conversion factor, this becomes $P = 3.34\,(1.34) = 4.48$ horsepower. Incidentally, the total propulsive force exerted by our 60 kg runner is $F_T = 520$ newtons = 117 pounds.

6. Of course, we could have computed the power directly from the quantity σ, which represents the runner's power input due to oxygen replenishment. So we have $P = m\sigma = (60\ \text{kg})\ (55.6\ \text{joules/kg s}) = 3{,}335$ joules/s.

7. It turns out that runners cannot maintain the velocity, $U_c = \sqrt{\sigma\tau}$ for an indefinite period of time. For distances greater than approximately $L = 10{,}000$ m, the velocity U begins to decrease markedly from the equation (24.17) value. Woodside (1991) suggests that muscle fatigue is the major reason for such decrease. He makes the assumption that the rate of energy loss due to fatigue at time, t, is proportional to the energy already utilized up to time, t. Woodside alters the energy relationship of equation (24.15) to take this fatigue effect into account. The outcome is the following modification of equation (24.16):

$$U = \sqrt{\frac{E_0\tau/T + \sigma\tau}{1 + \gamma T}}, \qquad (24.18)$$

where γ is a so-called fatigue constant; its numerical value is listed in table 24.3.

8. In figure 24.5, for men's and women's plots, equation (24.18) is shown as the extension of the running curve from a distance of about $L = 10{,}000$ m to $L = 100{,}000$ m.

Racing Records in the Future

As we conclude our brief study of running and racing, we look ahead and ask: How quickly will the present world records be broken and by how much? Of course, it is impossible to give precise answers to these questions. However, in 1982 the noted British physiologist B. B. Lloyd carried out studies concerning future world records of running races. His projections for the year 2000, along with the present records, are listed in table 24.4.

How Fast Can Swimmers Swim?

Over the years, sprinting and distance running have received far more attention in physiological studies than the numerous other types of athletic racing such as swimming, skating, and bicycle riding. This is not surprising when we consider the truly universal nature and role of running in the everyday lives of

TABLE 24.4

Present (1996) and projected (2000) record times of running races of men and women

	Men		*Women*	
Length L, m	*Present 1996 hr:min:sec*	*Projected 2000 hr:min:sec*	*Present 1996 hr:min:sec*	*Projected 2000 hr:min:sec*
100	9.84	9.82	10.49	10.77
200	19.34	19.63	21.34	21.35
400	43.29	42.60	47.60	45.49
800	1:41.73	1:38.80	1:53.28	1:44.42
1,500	3:28.56	3:25.90	3:50.46	3:42.13
1,609	3:44.39	3:42.70	4:15.61	4:00.94
3,000	7:28.96	7:20.70	8:06.11	8:02.27
5,000	12:56.96	12:36.90	14:37.33	13:52.84
10,000	26:58.38	26:03.80	29:31.78	28:46.31
42,195	2:06:50	2:02:21	2:21:06	2:14:37

Source: Data from Lloyd (1982).

humans. Even so, there is sufficient information concerning other forms of athletic competition to enable us to carry out mathematical analyses similar to those we did for running.

For example, let us examine the topic of freestyle swimming. The list of current records for men's races is presented in table 24.5 and for women's races in table 24.6. The information from both tables is displayed in graphical form in figure 24.6.

As before, the data shown in figure 24.6 are correlated by the linear relationship of equation (24.1):

$$\log_{10} L = \log_{10} a + c \log_{10} T,$$

which we can express in the form of equation (24.2):

$$L = aT^c.$$

TABLE 24.5

Records for freestyle swimming — men's races

Distance *L, m*	*Nominal time, min:sec*	*Time* *T, s*	*Velocity* $U = L/T, m/s$
50	21.81	21.81	2.29
100	48.42	48.42	2.07
200	1:46.69	106.69	1.87
400	3:45.00	225.00	1.78
800	7:47.85	467.85	1.71
1,500	14:43.48	883.48	1.70

Source: Data from Wright (1996).

TABLE 24.6

Records for freestyle swimming — women's races

Distance *L, m*	*Nominal time min:sec*	*Time* *T, s*	*Velocity* $U = L/T, m/s$
50	24.79	24.79	2.02
100	54.48	54.48	1.84
200	1:57.55	117.55	1.70
400	4:03.85	243.85	1.64
800	8:16.22	496.22	1.61
1,500	15:52.10	952.10	1.58

Source: Data from Wright (1996).

Using the freestyle swimming data listed in the tables and shown in figure 24.6, we determine that, for men, $a = 2.85$, $c = 0.918$, and, for women, $a = 2.41$, $c = 0.934$.

Finally, as we did for the running data, the so-called endurance equations are obtained. As before [equation (24.3)], these equa-

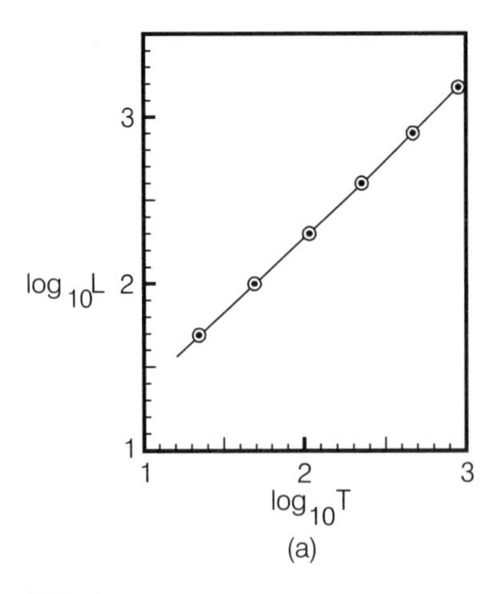

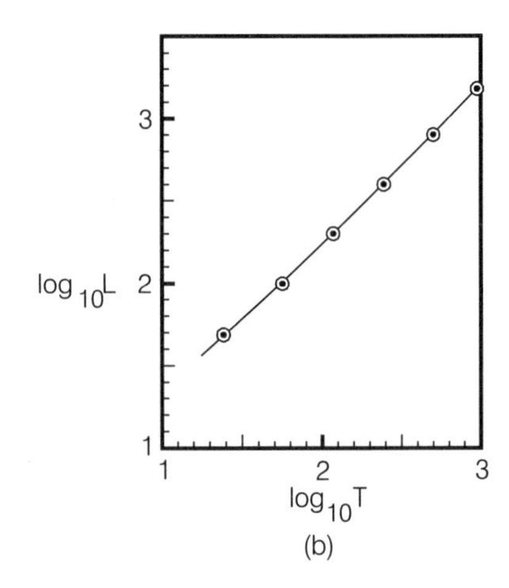

FIG. 24.6

Records for freestyle swimming races. Plots of distance, L, versus time, T, for (*a*) men's and (*b*) women's races.

tions have the form

$$U = K/L^n,$$

where $K = a^{1/c}$ and $n = (1/c) - 1$. For men, $K = 3.13$, $n = 0.089$, and for women, $K = 2.56$, $n = 0.071$. These curves are shown in figure 24.7.

We note that the endurance curves for freestyle swimming are similar in shape to those shown in figure 24.2 for the running data. It is interesting that the four values of the exponent n (for, respectively, running and swimming, men and women) are not greatly different. In contrast, the value of the constant K for the running data is substantially larger than the value of K for the swimming data.

With this information concerning freestyle swimming races, we could determine the numerical values of the various parameters shown in table 24.3 for the running races, including the maximum propulsion force and the maximum velocity.

In general, we note that for races of specified lengths L, the ratio of average running velocity to average swimming velocity ranges from around 5.0 for the shorter races to approximately 4.0

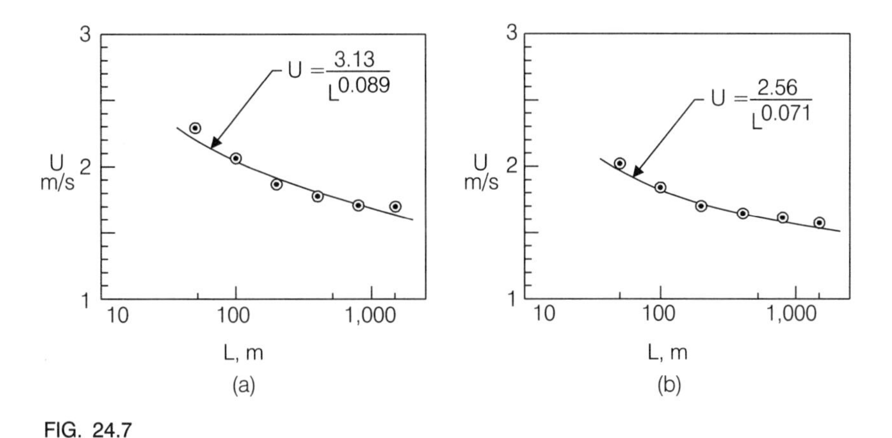

FIG. 24.7

Plots of endurance equations for freestyle swimming: (*a*) men's races and (*b*) women's races

for the longer. This says, in other words, that people can run about four or five times faster than they can swim.

PROBLEM. The simple model and method of analysis we devised for evaluating the running data serves as a suitable framework for an examination of the freestyle swimming data. Going a step further with a homework problem, select another type of racing and carry out a similar analysis. Some suggestions: cross-country skiing, cycling, rowing, and speed skating.

References

Abramowitz, M., and I. A. Stegun (1965). *Handbook of Mathematical Functions*. New York: Dover Publications.

Adair, R. K. (1990). *The Physics of Baseball*. New York: Harper and Row.

Adams, A. L. (1991). *Flying Buttresses, Entropy and O-Rings*. Cambridge, Mass.: Harvard University Press.

Alessandrini, S. M. (1995). "A motivational example for the numerical solution of two-point boundary-value problems." *SIAM Review* 37:423–427.

Allen, R. D. G. (1965). *Mathematical Economics*. 2nd ed. New York: Macmillan.

Anderson, J. D. (1991). *Fundamentals of Aerodynamics*. New York: McGraw-Hill.

Ashton, W. D. (1966). *The Theory of Road Traffic Flow*. London: Methuen and Co.

Ballard, R. D. (1987). *The Discovery of the Titanic*. Toronto: Madison Press.

Banks, R. B. (1994). *Growth and Diffusion Phenomena: Mathematical Frameworks and Applications*. Heidelberg: Springer-Verlag.

Barenblatt, G. I. (1996). *Scaling, Self-Similarity and Intermediate Asymptotics*. Cambridge, U. K.: Cambridge University Press.

Bartholomew, D. J. (1981). *Stochastic Models in Social Science*. Chichester, U.K.: Ellis Horwood.

Bascom, W. (1980). *Waves and Beaches: The Dynamics of the Ocean Surface*. Garden City, N.Y.: Anchor Press.

Bate, R. R., R. D. Mueller, and J. E. White (1971). *Fundamentals of Astrodynamics*. New York: Dover Publications.

Bearman, P. W., and J. K. Harvey (1976). Golf ball aerodynamics. *Aeronautical Quarterly* 27:112–122.

Bellman, R., and K. L. Cooke (1963). *Differential-difference Equations*. New York: Academic Press.

Beltrami, E. (1987). *Mathematics for Dynamic Modeling*. New York: Academic Press.

Bennett, W. R. (1976). *Scientific and Engineering Problem-Solving*. Englewood Cliffs, N.J.: Prentice-Hall.

Blackwell, B. F., and G. E. Reis (1974). *Blade Shape for a Troposkein Type of Vertical-Axis Wind Turbine*. Energy Report SLA-74-0154, Sandia National Laboratories, Albuquerque, N. Mex.

Bogin, B. (1988). *Patterns of Human Growth*. Cambridge, U.K.: Cambridge University Press.

Bolt, B. A. (1993). *Earthquakes*. New York: W. H. Freeman.

Borsellino, A., and V. Torre (1974). Limits to growth from Volterra theory of population. *Kybernetik* 16:113–118.

Brancazio, P. J. (1984). *Sports Science: Physical Laws and Optimal Performance*. New York: Simon and Schuster.

Brandt, S. (1976). *Statistical and Computational Methods in Data Analysis*. Amsterdam: North-Holland.

Brown, R. (1983). *Voyage of the Iceberg: The Story of the Iceberg that Sank the Titanic*. New York: Beaufort Books.

Bullard, F. M. (1976). *Volcanoes of the Earth*. Austin, Tx.: University of Texas Press.

Bullen, E. E. (1965). *An Introduction to the Theory of Mechanics*. London: Cambridge University Press.

Cooper, R. A., and A. J. Weekes (1983). *Data, Models and Statistical Analysis*. Oxford: Philip Allen Publishers.

Davis, P. J. (1961). *The Lore of Large Numbers*. New York: Random House.

Decker, R., and B. Decker (1981). "The eruptions of Mount St. Helens." *Scientific American* 244(3):68–80.

de Mestre, N. (1990). *The Mathematics of Projectiles in Sport*. Cambridge, U.K.: Cambridge University Press.

Dobson, A. J. (1983). *Introduction to Statistical Modeling*. London: Chapman and Hall.

Eaton, J. P., and C. A. Haas (1987). *Titanic: Destination Disaster*. New York: W. W. Norton.

Epstein, S. H. (1984). "Skiing on air." In *Newton at the Bat*, ed. F. W. Schrier, 134–139. New York: Scribner's.

Feller, W. (1940), "On the logistic law of growth and its experimental verifications in biology." Chapter 4, in *Applicable Mathematics of Nonphysical Phenomena*, eds. F. Olivera-Pinto and B. W. Connolly, 123–138. Chichester, U.K.: Ellis Horwood.

——— (1968). *An Introduction to Probability Theory and Its Applications*. New York: John Wiley.

Ferris, T., ed. (1991). *The World Treasury of Physics, Astronomy and Mathematics*. Boston: Little, Brown and Co.

Flora, S. D. (1956). *Hailstones of the United States*. Norman: University of Oklahoma Press.

Forrester, J. W. (1971). *World Dynamics*. Cambridge, Mass.: Wright-Allen Press.

French, A. P. (1971a). *Vibrations and Waves*. New York: W. W. Norton.

——— (1971b). *Newtonian Mechanics*. New York: W. W. Norton.

Gellert, W., Küstner, M. Hellwich, and H. Kästner (1977). *The VNR Concise Encyclopedia of Mathematics*. New York: Van Nostrand Reinhold.

Grieve, R. A. F. (1987). Terrestrial impact structures. *Annual Review of Earth and Planetary Sciences* 15:245–270.

——— (1990). "Impact cratering on the earth." *Scientific American* 262(4):44–51.

Haberman, R. (1977). *Mathematical Models*. Englewood Cliffs, N. J.: Prentice-Hall.

Haight, F. A. (1963). *Mathematical Theories of Traffic Flow*. New York: Academic Press.

Hamburger, H. S., ed. (1982). *Transportation and Traffic Engineering Handbook*. Englewood Cliffs, N.J.: Prentice-Hall.

Hammersley, J. M., and D. C. Handscomb (1964). *Monte Carlo Methods*. London: Methuen and Co.

Hart, D., and T. Croft (1988). *Modelling with Projectiles*. Chicester, U.K.: Ellis Horwood.

Hastings, N. A. J., and J. B. Peacock (1974). *Statistical Distributions*. New York: John Wiley.

Heilbroner, R. L. (1980). *The Worldly Philosophers*. 5th ed. New York: Simon and Schuster.

Heizer, R. T. (1978). "Energy and fresh water production from icebergs." In *Iceberg Utilization*, ed. A. A. Husseiny, 657–73. New York: Pergamon Press.

Hill, A. V. (1927). *Muscular Movement in Man: the Factors Governing Speed and Recovery from Fatigue*. New York: McGraw-Hill.

Hoerner, S. F. (1965). *Fluid-Dynamic Drag*. Bricktown, N. J.: Hoerner Fluid Dynamics.

Huff, D. (1954). *How to Lie with Statistics*. New York: W. W. Norton.

Husseiny, A. A., ed. (1978). *Iceberg Utilization*. New York: Pergamon Press.

Ipsen, D. C. (1960). *Units, Dimensions and Dimensionless Numbers*. New York: McGraw-Hill.

Isenberg, C. (1992). *The Science of Soap Films and Soap Bubbles*. New York: Dover Publications.

Jorgensen, T. P. (1994). *The Physics of Golf*. Woodbury, N. Y.: American Institute of Physics.

Kirby, R. S., S. Withington, A. B. Darling, and F. G. Kilgour (1990). *Engineering in History*. New York: Dover Publications.

Knight, C., and N. Knight (1971). "Hailstones." *Scientific American*. 224(3):97–103.

Knott, C. G. (1911). *Life and Scientific Work of Peter Guthrie Tait*. London: Cambridge University Press.

Krylov, I. A., and L. P. Remizov (1974). "Problem of the optimum ski jump." *PMM* (*Journal of Applied Mathematics, Moscow*) 38:717–20.

Keller, J. B. (1973). "A theory of competitive running." *Physics Today* 26:42–7.

——— (1974). "Optimal velocity in a race." *American Mathematical Monthly* 81:474–80.

Langhaar, H. L. (1951). *Dimensional Analysis and Theory of Models*. New York: John Wiley.

Leacock, S. B. (1956). "Mathematics for golfers." In *The World of Mathematics IV*, ed. J. R. Newman, 2456–59. New York: Simon and Schuster.

Le Méhauté, B. (1976). *Introduction to Hydrodynamics and Water Waves*. New York: Springer-Verlag.

Lide, D. R., ed. (1994). *CRC Handbook of Chemistry and Physics*. Boca Raton, Fla.: CRC Press.

Lighthill, M. J., ed. (1978). *Newer Uses of Mathematics*. Harmondsworth, U.K.: Penguin Books.

Lighthill, M. J., and G. B. Whitham (1955). "On kinematic waves. II. A theory of traffic flow on long crowded highways." *Proceedings of the Royal Society of London A* 229:317–45.

Lind, D. A., and S. P. Sanders (1996). *The Physics of Skiing*. Woodbury, N.Y.: American Institute of Physics Press.

Lloyd, B. B. (1982). "Athletic achievement—trends and limits." In *Science and Sporting Performance*, eds. D. Bruce and G. Thomas. Oxford: Clarendon Press.

Lock, G. S. H. (1990). *The Growth and Decay of Ice*. Cambridge, U.K.: Cambridge University Press.

Lord, W. (1955). *A Night to Remember*. New York: Holt, Rinehart and Winston.

Lorensen, W. E., and B. Yamrom (1992). "Green grass visualization." *IEEE Computer Graphics and Applications* 12(4):35–44.

Lotka, A. J. (1956). *Elements of Mathematical Biology*. New York: Dover Publications.

Loyrette, H. (1985). *Gustave Eiffel*. New York: Rizzoli International Publications.

MacDonald, N. (1989). *Biological Decay Systems: Linear Stability Theory*. Cambridge, U.K.: Cambridge University Press.

Maloney, R. (1956). "The inflexible logic." In *The World of Mathematics IV*, ed. J. R. Newman, 2262–67. New York: Simon Schuster.

Malthus, T. R. ([1798] 1970). *An Essay on the Principle of Population Growth*. Reprint, Harmondsworth, U.K.: Penguin Books.

McLachlan, B. G., G. Beaupre, A. B. Cox, and L. Gore (1983). "Falling dominoes." *SIAM Review* 25:403–4.

Meadows, D. L. (1974). *The Limits to Growth*. London: Pan Books.

Mehta, R. D. (1985). "Aerodynamics of sports balls." *Annual Review of Fluid Mechanics* 17:151–89.

Melosh, H. J. (1989). *Impact Cratering: A Geologic Process*. Oxford: Oxford University Press.

Mohazzabi, P., and J. H. Shea (1996). "High-altitude free fall." *American Journal of Physics* 64(10): 1242–46.

Moroney, M. J. (1974). *Facts from Figures*. 3rd ed. New York: Penguin Books.

Mosteller, F. (1965). *Fifty Challenging Problems in Probability with Solutions*. New York: Dover Publications.

Mosteller, F., S. E. Fienberg, and R. E. K. Rourke (1983). *Beginning Statistics with Data Analysis*. Reading, Mass.: Addison-Wesley.

Myles, D. (1985). *The Great Waves*. New York: McGraw-Hill.

Neft, D. S., R. M. Cohen, and R. Korch (1992). *The Sports Encyclopedia: Pro Football*. 10th ed. New York: St. Martin's Press.

Newman, J. R., ed. (1956). *The World of Mathematics*. New York: Simon and Schuster.

Newton, R. E. I. (1990). *Wave Physics*. London: Edward Arnold.

Perelman, Y. (1979). *Mathematics Can Be Fun*. Moscow: Mir Publishers.

Perlman, E. (1984). "Ballistics of speed skiing." In *Newton at the Bat*. ed. F. W. Schrier, 131–33. New York: Scribner's.

Péronnet, F., and G. Thibault (1989). "Mathematical analysis of running performance and world running records." *Journal of Applied Physiology* 67:453–65.

Prigogine, I., and R. Herman (1971). *A Kinetic Theory of Vehicular Traffic*. New York: American Elsevier.

Pritchard, W. G. (1993). "Mathematical models of running." *SIAM Review* 35: 359–79.

Pritchard, W. G., and J. K. Pritchard (1994). "Mathematical models of running." *The American Scientist* 82:546–53.

Raine, A. E. (1970). "Aerodynamics of skiing." *Science Journal* 6:26–39.

Rastrigin, L. (1984). *This Chancy, Chancy, Chancy World*. Moscow: Mir Publishers.

Rhee, H. K., R. Aris, and N. R. Amundson (1986). *First-Order Partial Differential Equations*, Vol. I. Englewood Cliffs, N.J.: Prentice-Hall.

Roos, D. S. (1972). "A giant hailstone from Kansas in free fall." *Journal of Applied Meteorology* II:1008–11.

Rozanov, Y. A. (1977). *Probability Theory: A Concise Course*. New York: Dover Publications.

Saarinen, A. B., ed. (1962). *Eero Saarinen on His Work*. New Haven, Conn.: Yale University Press.

Saaty, T. L. (1981). *Modern Nonlinear Equations*. New York: Dover Publications.

Scudo, F. M., and J. P. Ziegler, eds. (1978). *The Golden Age of Theoretical Ecology: 1923–1940*. Lecture Notes in Biomathematics, no. 22. Berlin: Springer-Verlag.

Sellers, G. R., S. B. Vardeman, and A. F. Hackert (1992). *A First Course on Statistics*. New York: Harper Collins.

Shaw, E. E. (1978). "Mechanics of a chain of dominoes." *American Journal of Physics* 46:640–42.

Shoemaker, E. M. (1983). "Asteroid and comet bombardment of the earth." *Annual Review of Earth and Planetary Sciences* 11:461–94.

Simkin, T., and R. S. Fiske (1983). *Krakatau* 1883. Washington D.C.: Smithsonian Institution Press.

Stonehouse, B. (1990). *North Pole, South Pole*. London: Prion.

Streeter, V. L., and E. B. Wylie (1985). *Fluid Mechanics*. 8th ed. New York: McGraw-Hill.

Tait, P. G. (1890, 1891, 1893). "Some points in the physics of golf." *Nature* 42:420–23, 44:497–98, 48:202–4.

Tanner, J. M. (1978). *Foetus into Man: Physical Growth from Conception to Maturity*. London: Open Books Publishing.

Tanner, J. M., and G. R. Taylor (1981). *Growth*. New York: Time-Life Books.
Thompson, D. W. (1963). *On Growth and Form*. 2nd ed. Cambridge, U.K.: Cambridge University Press.
——— (1986). *Introduction to Space Dynamics*. New York: Dover Publications.
Townend, M. S. (1984). *Mathematics in Sport*. Chicester, U.K.: Ellis Horwood.
Tricker, R. A. R. (1964). *Bores, Breakers, Waves and Wakes: An Introduction to the Study of Waves on Water*. London: Mills and Boon.
U.S. Bureau of the Census (1994). *Statistical Abstract of the United States*. Washington D.C.: Government Printing Office.
Van Der Zee, J. (1986). *The Gate*. New York: Simon and Schuster.
Visualization Society of Japan (1993). *The Fantasy of Flow: The World of Fluid Flow Captured in Photographs*. Tokyo: Ohmsha Ltd.
Vonnegut, K. (1970). "Epicac." In *Welcome to the Monkey House*, 277–84. New York, Dell Publishing.
Wadhams, P. (1996). "The resource potential of Antarctic icebergs." In *Oceanography: Contemporary Readings in Ocean Sciences*. ed. R. G. Pirie, 358–73. New York: Oxford University Press.
Ward-Smith, A. J. (1985). "A mathematical theory of running based on the first law of thermodynamics." *Journal of Biomechanics* 18:337–49.
Ward-Smith, A. J., and D. Clements (1982). "Experimental determination of the aerodynamic characteristics of ski-jumpers." *Aeronautical Journal* 86: 384–91.
——— (1983). "Numerical evaluation of the flight mechanics and trajectory of a ski-jumper." *Acta Applicandae Mathematicae* 1:301–14.
Watts, R. G., and A. T. Bahill (1990). *Keep Your Eye on the Ball: The Science and Folklore of Baseball.* New York: W. H. Freeman.
Weaver, W. (1963). *Lady Luck. The Theory of Probability.* Garden City, N.Y.: Anchor Books.
Weeks, W. F., and M. Mellor (1978). "Some elements of iceberg technology." In *Iceberg Utilization*, ed. A. A. Husseiny, 45–98. New York: Pergamon Press.
Whitham, G. B. (1974). *Linear and Nonlinear Waves*. New York: John Wiley.
Williams, K. R. (1985). "Biomechanics of running." In *Exercise and Sports Sciences Reviews* 13, ed. R. L. Terjung, 389–441. New York: Macmillan.
Woodside, W. (1991). "The optimal strategy for running a race." *Mathematical and Computer Modelling* 15(10):1–12.
Wright, J. W., ed. (1996). *The Universal Almanac 1997*. Kansas City, Mo.: Andrews and McMeel.
Zwillinger, D. (1989). *Handbook of Differential Equations*. San Diego: Academic Press.

Index

ABOUT THE AUTHOR

Robert B. Banks is a former Professor of Engineering at Northwestern University and former Dean of Engineering at the University of Illinois at Chicago. He served with the Ford Foundation in Mexico City and with the Asian Institute of Technology in Bangkok. He has won numerous national and international honors, including being named Commander of the Order of the White Elephant by the King of Thailand and Commandeur dans l'Ordre des Palmes Academiques by the government of France. He is currently living in California and working on his next book.